COMPUTER SCIENCE, TECHNOLOGY AND APPLICATIONS

FEDERAL GEOSPATIAL INFORMATION MANAGEMENT AND COORDINATION

COMPUTER SCIENCE, TECHNOLOGY AND APPLICATIONS

Additional books in this series can be found on Nova's website under the Series tab.

Additional E-books in this series can be found on Nova's website under the E-book tab.

EARTH SCIENCES IN THE 21ST CENTURY

Additional books in this series can be found on Nova's website under the Series tab.

Additional E-books in this series can be found on Nova's website under the E-book tab.

COMPUTER SCIENCE, TECHNOLOGY AND APPLICATIONS

FEDERAL GEOSPATIAL INFORMATION

MANAGEMENT AND COORDINATION

ANDREW HALPERN
AND
MARK TROUPP
EDITORS

nova publishers
New York

Library of Congress Cataloging-in-Publication Data

ISBN 978-1-62257-572-5

Published by Nova Science Publishers, Inc. † New York

CONTENTS

PREFACE

Geospatial information is data referenced to a place--a set of geographic coordinates--which can often be gathered, manipulated, and displayed in real time. A Geographic Information System (GIS) is a computer data system capable of capturing, storing, analyzing, and displaying geographically referenced information. The federal government and policy makers increasingly use geospatial information and tools like GIS for producing floodplain maps, conducting the census, mapping foreclosures, congressional redistricting, and responding to natural hazards such as wildfires, earthquakes, and tsunamis. For policy makers, this type of analysis can greatly assist in clarifying complex problems that may involve local, state, and federal government, and affect businesses, residential areas, and federal installations. This book provides a primer on geospatial data and GIS and explores the complexity of managing, sharing, and using geospatial data from a variety of sources and across the breadth of the federal government.

Chapter 1 – Geospatial information is data referenced to a place—a set of geographic coordinates—which can often be gathered, manipulated, and displayed in real time. A Geographic Information System (GIS) is a computer data system capable of capturing, storing, analyzing, and displaying geographically referenced information. The federal government and policy makers increasingly use geospatial information and tools like GIS for producing floodplain maps, conducting the census, mapping foreclosures, congressional redistricting, and responding to natural hazards such as wildfires, earthquakes, and tsunamis. For policy makers, this type of analysis can greatly assist in clarifying complex problems that may involve local, state, and federal government, and affect businesses, residential areas, and federal installations.

Examples of how GIS and geospatial data are used within and outside the federal government are growing rapidly. In this chapter, a few examples are provided that describe the real-time or near real-time data analysis in the case of a California wildfire; policy analysis in support of a Base Realignment and Closure decision in Virginia Beach; and analysis of foreclosure patterns using census and other data for the New York City area. An additional example is provided demonstrating the burgeoning interaction of GIS and social media. In this case, Japanese citizens collected and provided census records, maps, and other information—a variant of "crowdsourcing"—to a GIS team. The team assembled the information into data layers supporting an interactive map to assist humanitarian organizations working in areas of Japan damaged by the March 11, 2011, earthquake and tsunami.

Office of Management and Budget (OMB) Circular A-16, first issued in 1953, gives direction for federal agencies that produce, maintain, or use geospatial data. OMB Circular A-16 has been revised and updated in 1967, 1990, and 2002. Most recently, the Obama Administration issued supplemental guidance to Circular A-16 that labeled federal geospatial data a capital asset and referred to its acquisition and management in terms analogous to financial assets. How well these "assets" are managed depends, in part, on how the federal government is structured to organize and coordinate its geospatial enterprise.

That structure is embodied in the Federal Geographic Data Committee (FGDC), comprising 10 cabinet-level departments and 9 independent agencies. OMB Circular A-16, via its revisions and supplemental guidance, as well as Executive Order 12906, issued in 2004, gives the FGDC primary responsibility for developing the National Spatial Data Infrastructure (NSDI). The NSDI can be thought of as the infrastructure for federal geospatial "assets," or the means by which federal geospatial data are acquired, processed, distributed, used, maintained, and preserved.

The 112th Congress in its oversight role may have an interest in the programs and geospatial assets belonging to most federal departments and agencies within the framework of the NSDI. This chapter describes some of these programs to give a sense of the breadth and complexity of the federal geospatial enterprise.

Chapter 2 – Congress has recognized the challenge of coordinating and sharing geospatial data from the local, county, and state level to the national level, and vice versa. The cost to the federal government of gathering and coordinating geospatial information has also been an ongoing concern. As much as 80% of government information has a geospatial component, according to various sources.

The federal government's role has changed from being a primary provider of authoritative geospatial information to coordinating and managing geospatial data and facilitating partnerships. Congress explored issues of cost, duplication of effort, and coordination of geospatial information in hearings during the 108th Congress. However, challenges to coordinating how geospatial data are acquired and used—collecting duplicative data sets, for example—at the local, state, and federal levels, in collaboration with the private sector, are not yet resolved. Two bills introduced in the 112th Congress, H.R. 1620 and H.R. 4322, would address aspects of duplication and coordination of geospatial information.

The federal government has recognized the need to organize and coordinate the collection and management of geospatial data since at least 1990, when the Office of Management and Budget (OMB) revised Circular A-16 to establish the Federal Geographic Data Committee (FGDC) and to promote the coordinated use, sharing, and dissemination of geospatial data nationwide. OMB Circular A-16 also called for development of a national digital spatial information resource to enable the sharing and transfer of spatial data between users and producers, linked by criteria and standards. Executive Order 12906, issued in 1994, strengthened and enhanced Circular A-16, and specified that FGDC shall coordinate development of the National Spatial Data Infrastructure (NSDI). On November 10, 2010, OMB issued supplemental guidance to Circular A-16 that labels geospatial data as a "capital asset," and refers to its acquisition and management in terms analogous to financial assets to be managed as a National Geospatial Data Asset Portfolio. It will likely take some time, and several budget cycles, to track whether agencies are adhering to the "portfolio-centric model" of geospatial data management outlined in the supplemental guidance. The 112th Congress may examine its oversight role in the implementation of OMB Circular A-16, particularly in how federal agencies are coordinating their programs that have geospatial assets.

The high-level leadership and broad membership of the FGDC—10 cabinet-level departments and 9 other federal agencies—highlights the importance of geospatial information to the federal government. Questions remain, however, about how effectively the FGDC is fulfilling its mission. Has this organizational structure worked? Can the federal government account for the costs of acquiring, coordinating, and managing geospatial information? How well is the federal government coordinating with the state and local entities that have an increasing stake in geospatial information? What is the role of the private sector?

State-level geospatial entities, through the National State Geographic Information Council, also embrace the need for better coordination. However, the states are sensitive to possible federal encroachment on their prerogatives to customize NSDI to meet the needs of the states. In early 2009, several proposals were released calling for efforts to create a national Geospatial Information System (GIS). Language in the proposals attempted to make the case for considering such efforts part of the national investment in critical infrastructure. Congress may consider how a national GIS or geospatial infrastructure would be conceived, perhaps drawing on proposals for these national efforts and how they would be similar to or differ from current efforts.

Chapter 3 – We are pleased to share version 4 of the Modernization Roadmap for the Geospatial Platform (Roadmap). This final version of the Roadmap follows a stakeholder engagement process intended to share the vision for the Platform and gather feedback from the geospatial community regarding concepts discussed in the Roadmap. Feedback on the Roadmap to date has been positive, while at the same time identifying some challenging issues and questions that will be addressed as implementation occurs.

Developing the Roadmap has been an evolving process that began in February 2010 with the release of the President‘s 2011 Budget, which provided initial direction to develop the Geospatial Platform. An earlier draft Roadmap (version 3), was approved by the Office of Management and Budget (OMB) on November 10, 2010. That document, prepared by member agencies of the Federal Geographic Data Committee (FGDC), received broad Federal review and upon OMB approval set the stage for a more robust stakeholder engagement process.

The geospatial community at large has been engaged in providing feedback on the Roadmap over the past several months through a variety of mechanisms that include:

- Review and feedback from the National Geospatial Advisory Committee (NGAC), a Federal Advisory Committee that provides advice to the Federal government on geospatial matters;
- Presentations and discussions at geospatial conferences and related professional meetings; and
- Development of an IdeaScale Web-based forum for gathering input and feedback through the Geospatial Platform Web page_(http://www.geoplatform.gov).

These outreach mechanisms resulted in valuable feedback that can be broadly categorized into questions and comments related to the following areas:

- Policy and management;
- Governance;
- Lessons to be learned from other model programs, and;
- Clarification of concepts.

The Geospatial Platform team has listened to the insights, ideas, and concerns expressed by geospatial stakeholders and offer the following summary of the approach for addressing each these areas of feedback as the Geospatial Platform moves into implementation throughout 2011 and beyond. This final Roadmap document does not fully address all of the issues identified through the stakeholder engagement process. Some of the issues raised through the outreach process will be addressed through the implementation process for the Platform.

Chapter 4 –Over the past year, the partner agencies of the FGDC have made significant progress on a number of fronts. For example, the FGDC has made significant progress in conceptualizing and beginning implementation of the Geospatial Platform initiative. The Geospatial Platform, which was described in the President's fiscal year 2011 budget, will be an accessible, reliable, and cost-effective collection of data, applications, and services that will provide a foundation to help address the White House's priorities for data sharing, transparency, and collaboration. In 2011, the FGDC developed a conceptual roadmap for the Platform and received very helpful feedback from external partners on the National Geospatial Advisory Committee and through a public outreach process. The FGDC also began the implementation of the first operational phase of the Platform in 2011. We are excited by the prospects for this new initiative, and we look forward to working with our partners in the coming year to realize the vision for the Platform.

The FGDC has also provided support for key White House initiatives. For example, the FGDC worked with the Office of Management and Budget (OMB) to develop new Supplemental Guidance to OMB Circular A–16, "Coordination of Geographic Information and Related Spatial Data Activities." The new guidance outlines a process to use portfolio management approaches to more effectively manage Federal geospatial assets and investments. OMB issued the new guidance to agencies in November 2010.

In addition, the FGDC has made significant progress in the area of geospatial standards. The FGDC adopted 64 external standards, including many from the International Organization for Standardization (ISO) and the Open Geospatial Consortium, and also adopted the U.S. Thoroughfare, Landmark, and Postal Address Data Standard. In addition, the FGDC is coordinating standardization activities with the Department of Defense. Standards are a key component of the National Spatial Data Infrastructure and we are pleased with the progress in this vital area.

Fiscal year 2012 will focus on continued collaboration to better enable public-private partnerships and investments, and continued progress on development of the Geospatial Platform and other tools and data for the benefit of the agencies and public we serve. We are also rejuvenating the FGDC Steering Committee and working with senior officials in our partner agencies to advance national geospatial programs.

In: Federal Geospatial Information
Editors: Andrew Halpern and Mark Troupp

ISBN 978-1-62257-572-5

Chapter 1

GEOSPATIAL INFORMATION AND GEOGRAPHIC INFORMATION SYSTEMS (GIS): AN OVERVIEW FOR CONGRESS*

Peter Folger

SUMMARY

Geospatial information is data referenced to a place—a set of geographic coordinates—which can often be gathered, manipulated, and displayed in real time. A Geographic Information System (GIS) is a computer data system capable of capturing, storing, analyzing, and displaying geographically referenced information. The federal government and policy makers increasingly use geospatial information and tools like GIS for producing floodplain maps, conducting the census, mapping foreclosures, congressional redistricting, and responding to natural hazards such as wildfires, earthquakes, and tsunamis. For policy makers, this type of analysis can greatly assist in clarifying complex problems that may involve local, state, and federal government, and affect businesses, residential areas, and federal installations.

Examples of how GIS and geospatial data are used within and outside the federal government are growing rapidly. In this chapter, a few examples are provided that describe the real-time or near real-time data analysis in the case of a California wildfire; policy analysis in support of a Base Realignment and Closure decision in Virginia Beach; and analysis of foreclosure patterns using census and other data for the New York City area. An additional example is provided demonstrating the burgeoning interaction of GIS and social media. In this case, Japanese citizens collected and provided census records, maps, and other information—a variant of "crowdsourcing"—to a GIS team. The team assembled the information into data layers supporting an interactive map to assist humanitarian organizations working in areas of Japan damaged by the March 11, 2011, earthquake and tsunami.

Office of Management and Budget (OMB) Circular A-16, first issued in 1953, gives direction for federal agencies that produce, maintain, or use geospatial data. OMB

---

* This is an edited, reformatted and augmented version of a Congressional Research Service publication, CRS Report for Congress R41825, prepared for Members and Committees of Congress, from www.crs.gov, dated May 18, 2011.

Circular A-16 has been revised and updated in 1967, 1990, and 2002. Most recently, the Obama Administration issued supplemental guidance to Circular A-16 that labeled federal geospatial data a capital asset and referred to its acquisition and management in terms analogous to financial assets. How well these "assets" are managed depends, in part, on how the federal government is structured to organize and coordinate its geospatial enterprise.

That structure is embodied in the Federal Geographic Data Committee (FGDC), comprising 10 cabinet-level departments and 9 independent agencies. OMB Circular A-16, via its revisions and supplemental guidance, as well as Executive Order 12906, issued in 2004, gives the FGDC primary responsibility for developing the National Spatial Data Infrastructure (NSDI). The NSDI can be thought of as the infrastructure for federal geospatial "assets," or the means by which federal geospatial data are acquired, processed, distributed, used, maintained, and preserved.

The 112th Congress in its oversight role may have an interest in the programs and geospatial assets belonging to most federal departments and agencies within the framework of the NSDI. This chapter describes some of these programs to give a sense of the breadth and complexity of the federal geospatial enterprise.

INTRODUCTION

The explosion of consumer demand for geospatial information and tools such as geographic information systems (GIS) to manipulate and graphically display such information has brought GIS into the daily lives of millions of Americans, whether they know it or not. Google Earth and handheld or dashboard navigation systems represent enormously popular examples of the wide variety of applications made possible through the availability of geospatial information.[1] The release of Google Earth in 2005 represented a paradigm shift in the way people understand geospatial information, according to some observers, because it offered multi-scale visualization of places and locations around the globe that was free and easy to use.[2]

This chapter provides a primer on geospatial data and GIS and provides several examples of their use.

It should be recognized that the amount of geospatial data is expanding rapidly, the methods for acquiring geospatial data are growing, and the ways geospatial data are being used are diversifying throughout local and state governments, as well as within the federal government.

It is beyond the scope of this chapter to present the universe of geospatial data and its utility to the federal government. However, the federal government has had and continues to have a major role in the overall framework for geospatial data, including its organization, coordination, and sharing among federal agencies and with state and local entities.

In addition to providing basic information on GIS and geospatial information, this chapter describes the federal geospatial enterprise and how it is organized. Given the complexity of managing, sharing, and using geospatial data from a variety of sources and across the breadth of the federal government, the 112th Congress in its oversight role may have an interest in the programs and geospatial "assets" belonging to most federal departments and agencies.[3]

A discussion of classified geospatial information and national security issues is beyond the scope of this chapter.

GIS AND GEOSPATIAL DATA: THE BASICS

GIS is a computer data system capable of capturing, storing, analyzing, and displaying geographically referenced information—information attached to a location, such as latitude and longitude, or street location.[4]

Geographically referenced information is also known as geospatial information. Types of geospatial information include features like highway intersections, office buildings, rivers, the path of a tornado, the San Andreas Fault, or congressional district boundaries. Information associated with a specific location is referred to in GIS parlance as an attribute,[5] such as the population of a congressional district, or amount of movement per year along the San Andreas Fault. Other terms common to geospatial data and GIS analysis are described in the box below.

The power of GIS is the ability to combine geospatial information in unique ways—by layers or themes—and extract something new. For instance, a GIS analysis might include the location of a highway intersection and the average number of vehicles that flow through the intersection throughout the day, and extract information useful for locating a business. GIS might include both the location of a river and the water depth along its course by season, and enable an analysis of the effects of development on runoff within the watershed.

Overlaying the path of a severe thunderstorm with geospatial data on the types of structures encountered—homes, stores, schools, post offices—could inform an analysis of what types of building construction can survive high winds and hail.

Sources and Types of Geospatial Data

Geospatial data may be acquired by federal, state, tribal, county, and local governments, private companies, academic institutions, and nonprofit organizations. The collection and management of geospatial data are considered by many to be the costliest components of a GIS—some experts attribute close to 80% of GIS total costs to data acquisition.[6]

Geospatial data can be acquired using a variety of technologies. Land surveyors, census takers, aerial photographers, police, and even average citizens with a GPS-enabled cell phone can collect geospatial data using GPS or street addresses that can be entered into GIS.[7]

The attributes of the collected data, such as land-use information, demographics, landscape features, or crime scene observations, can be entered manually or, in the case of a land survey map, digitized from a map format to a digital format by electronic scanning. Remote sensing data from satellites is acquired digitally and communicated to central facilities for processing and analysis in GIS.

Digital satellite images, for example, can be analyzed in GIS to produce maps of land cover and land use. When different types of geospatial data are combined in GIS (e.g., through combining satellite remote sensing land use information with aerial photograph data on housing development growth), the data must be transformed so they fit the same coordinates. GIS uses the processing power of a computer, together with geographic mapping techniques (cartography), to transform data from different sources onto one projection and one scale so that the data can be analyzed together.

Geospatial and GIS Terminology

Attribute: descriptive information about the properties of events, features, or entities associated with a location, such as the ownership of a parcel of land, or the population of a neighborhood, or the wind speed and direction over a point on the ground.

Bathymetry: the science of measuring and charting the depths of water bodies to determine the topography of a lake bed or seafloor.

Cadastre: the map of ownership and boundaries of land parcels, often used to record ownership and assist in calculating taxes.

Cartography: the study and practice of making maps.

Datum: a definition of the origin, orientation, and scale of the coordinate system; usually a system of coordinate positions on a surface (horizontal datum) or heights above or below a surface (vertical datum).

Geocoding: assignment of alphanumeric codes or coordinates to geographically referenced data. Examples include the two-letter country codes, or the coordinates of a residence computed from its address. The end result is spatial data that can be displayed as features on a map.

Geodetic control: horizontal or vertical survey monuments that are primarily intended to serve as reference positions for other surveys or that serve to extend the national geodetic control networks.

Geographic information system (GIS): a digital database in which information is stored by its spatial coordinate system, which allows for data input, storage, retrieval, management, transformation, analysis, reporting, and other activities. GIS is often envisioned as a process as much as a physical entity for data.

Geospatial data: information that identifies the geographic location and characteristics of natural and constructed features and boundaries on Earth.

Global Positioning System (GPS): a navigation system supported by a constellation of 24 satellites owned and operated by the U.S. Department of Defense. The satellites transmit precise microwave signals that enable GPS receivers such as handheld devices or receivers installed in automobiles to determine their location, speed, and direction.

Hydrography: the charting and description of bodies of water.

LIDAR: acronym for Light Detection and Ranging, a remote sensing technique that uses laser pulses to determine elevation with high accuracy, usually from an aerial survey.

Map: a two-dimensional visual portrayal of geospatial data. The map is not the data itself.

Metadata: information about the quality, content, condition, and other characteristics of data. It may describe and document how, when, where, and by whom the data was collected, among other types of information.

Orthoimagery: An aerial photograph or image from which distortions resulting from camera tilt and ground relief have been removed. An orthophotograph or orthoimage has a uniform scale and can be used as a map.

Polygon: a feature in GIS used to represent areas (versus a point, or a line). A polygon is defined by the lines that make up its boundary. On a map, the closed shape representing the area is defined by a connected sequence of coordinates, or x,y pairs.

Projection: a mathematical means of transferring information from the Earth's three-dimensional, curved surface onto a two-dimensional map or computer screen.

Sources: Environmental Systems Research Institute, Inc., (ESRI), GIS Dictionary, http://support.esri.com/en/ knowledgebase/Gisdictionary/browse; Urban and Regional Information Systems Association (URISA), GIS Glossary of Terms, http://www.urisa.org/files/publications/gis_glossary/gis_glossary.pdf.

Geospatial Data from Local, State, and Federal Governments and the Private Sector

Local and state governments provide geospatial data for use in GIS for a variety of public services such as land records, property taxation, local planning, subdivision control and zoning, and others.[8] Local governments often contract with private-sector companies to acquire more recent and higher-resolution data than what is available to the federal government.[9] Whether and how the most up-to-date and detailed geospatial information is made available to users other than the local government for which the data were acquired are long-standing issues. For example, in the immediate aftermath of a natural disaster, such as Hurricane Katrina in 2005, it may be important for the federal government to acquire the most current and detailed geospatial information about the disaster area. In many instances, however, impediments to data sharing such as lack of interoperability between systems, restrictions on use, concerns about data security, and a lack of knowledge about what data exist and where the data can be found could hinder a timely and effective emergency response.[10]

The federal government sometimes acquires geospatial data for federal needs, such as for updating floodplain maps from paper flood insurance rate maps to a digital format. Assessing the need to update floodplain maps on a periodic basis is required by law,[11] and the Federal Emergency Management Agency (FEMA) has spent over $1.4 billion since FY2003 to convert paper flood insurance rate maps (FIRMs) to digital flood insurance rate maps (DFIRMs) and to produce a format usable in GIS.[12] Simply converting paper maps to digital formats does not necessarily improve their accuracy, which often depends on the resolution of the original data. New techniques for collecting more data, such as Light Detection and Ranging (LIDAR), would help produce more accurate floodplain maps.[13]

Geospatial data are increasingly acquired and provided by the private sector, and many companies as well as professional organizations support and promote the role of private-sector data providers. One organization, the Management Association for Private Photogrammetric Surveyors (MAPPS), bills itself as the only national association exclusively composed of private geospatial firms.[14]

MAPPS itself is a member of a larger coalition—the Coalition of Geospatial Organizations (COGO). COGO is comprised of 11 organizations and associations involved in geospatial data and policy issues.[15]

GIS Layers or Themes

The attributes of different types of geospatial data—such as land ownership, roads and bridges, buildings, lakes and rivers, counties, or congressional districts—can each constitute a layer or theme in GIS. (See *Figure 1* for a schematic representation of data layers in GIS.)

GIS has the ability to link and integrate information from several different data layers or themes over the same geographic coordinates, which is very difficult to do with any other means. For example, GIS could combine a major road from one data layer as the boundary dividing land zoned for commercial development with the location of wetlands from another data layer. Precipitation data, from a third layer, could be combined with a fourth data layer that shows streams and rivers.

GIS could then be used to calculate where and how much runoff might flow from the commercial development into the wetlands. Thus, the power of GIS analysis can be used to create a new way to interpret information that would otherwise be very difficult to visualize and analyze.

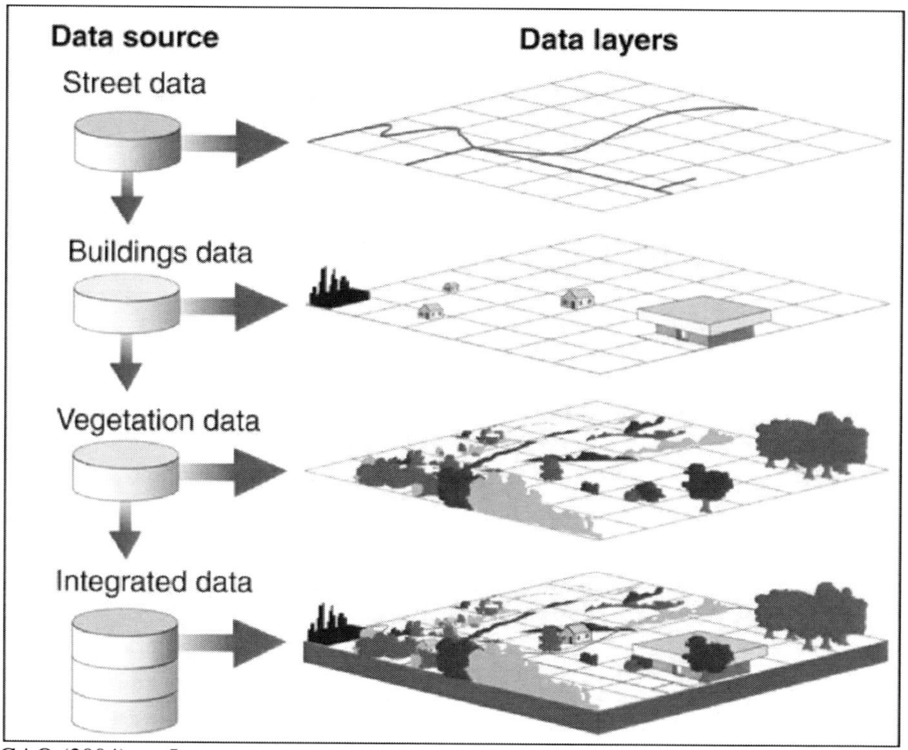

Source: GAO (2004), p. 5.

Figure 1. Example of GIS Data Layers or Themes.

EXAMPLES OF WHY AND HOW GEOSPATIAL INFORMATION IS USED

California Wildfires

Timeliness is an important factor for some uses of geospatial information. An example is the southern California wildfires during 2008. One of the worst fires in the region, known as the Sylmar fire, began on the evening of November 14, 2008. It forced the evacuation of 10,000 people and shut down major freeways near the town of Sylmar on the edge of the Angeles National Forest.[16] The speed of the fire's progress made it difficult to know where the fire was heading and to visualize escape routes. In addition, the fire jumped Interstate 210 (I-210) and Interstate 5 (I-5), two major routes of egress, on Saturday, November 15.

To assist in real-time decision making, the fire's progress was posted on the Internet in near real-time by several organizations, using reports from the ground, and the information about the fire was displayed on underlying street maps (showing where the fire crossed I-5 and I-210), terrain maps, and satellite images. (See *Figure 2.)* The Sylmar fire example underscores the informational power available when geospatial information is combined with tools for displaying the information, such as GIS and the Internet. In this instance, timeliness—the ability to post the geospatial information quickly—enhanced its value to the data users, citizens trying to avoid the path of the fire.

Although timeliness is often important, the analytical power resulting from combining geospatial information with GIS more typically underscores its value to policy makers at all levels. GIS often provides for unique analyses of disparate types of information—linked by their spatial coordinates—to help resolve policy questions. For policy makers, this type of analysis can greatly assist in clarifying complex problems that may involve local, state, and federal government, and may affect businesses, residential areas, and federal installations.

Base Realignment and Closure (BRAC) Program

The Base Realignment and Closure (BRAC) program is the process by which unneeded military facilities are identified and transferred to other federal agencies or disposed of. The City of Virginia Beach, VA, used GIS in its response to the 2005 BRAC Commission's recommendation to realign Naval Air Station Oceana, located near the population center of the city. The BRAC Commission was concerned that the city's land use was encroaching on the air station; in particular, the city was impinging on the noise zones and accident potential zone (APZ) around the air station. Because the recommended realignment of Oceana would likely cause Virginia Beach to suffer significant economic losses, the city sought to establish a baseline—using GIS— to understand the status of encroachment. In addition, the GIS analysis could inform city leaders about how to modify the municipal land use ordinance to prevent encroachment on the air station and forestall its realignment.

To establish a baseline, city planners and GIS analysts overlaid noise zones and APZ, property, land use, zoning, and other sets of geospatial data—known as attributed boundary layers—to determine current land use and development. Within the GIS analysis, these sets of geospatial data were joined with land parcel information, and with various external databases held within the planning, real estate assessor's, and commissioner's offices. By combining geospatial data with non-spatial data, the GIS analysts helped land planners determine how the land around the air station was being used, and therefore its compatibility with the Navy's requirements. (See *Figure 3.*) The GIS analysis also enabled the city to summarize property values and acreage by its use: undeveloped, commercial, or residential. GIS helped the Virginia Beach city planners to identify on one map all of the land use around the air station (Figure 3). GIS analysts also provided a model of underdeveloped land—land that had additional existing by-right development capacity—but which if developed could exacerbate the encroachment problem for the Navy.

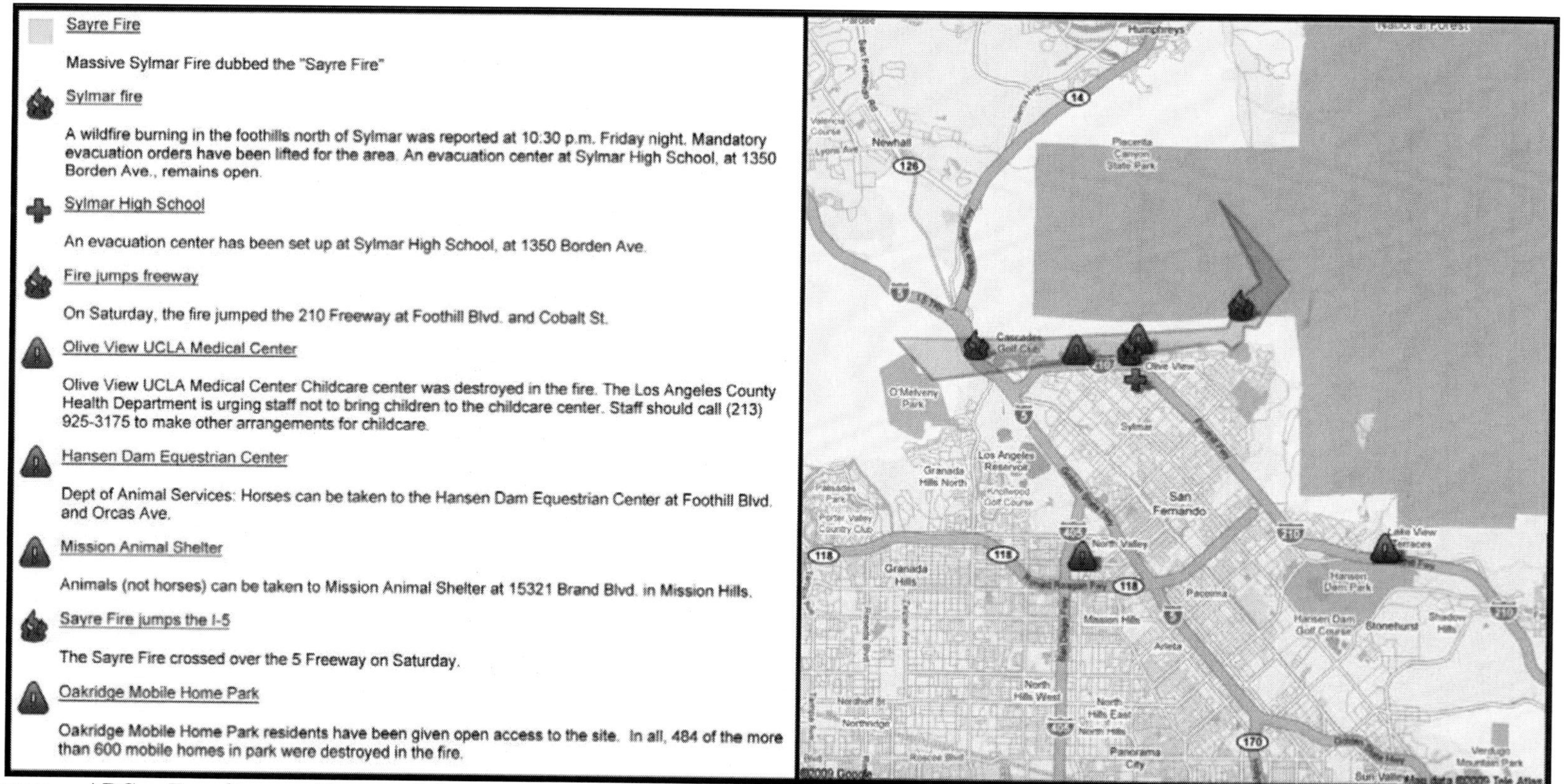

Source: ABC Eyewitness News, Sylmar Wildfire, created Nov. 15, 2008, updated Nov. 20, 2008. See http://maps.google 100866907082629170478.00045bb5e2170708e9258&t=h&source=embed&ll=34.314638,-118.436834&spn=0.251809,0.528717&z=12. Modified by CRS.

Notes: the path of the fire with the annotation is shown with an underlying street map. The original interactive website also allows the user to choose an underlying terrain map or satellite image map.

Figure 2. Snapshot of the Path of the 2008 Sylmar Fire Near Los Angeles, CA.

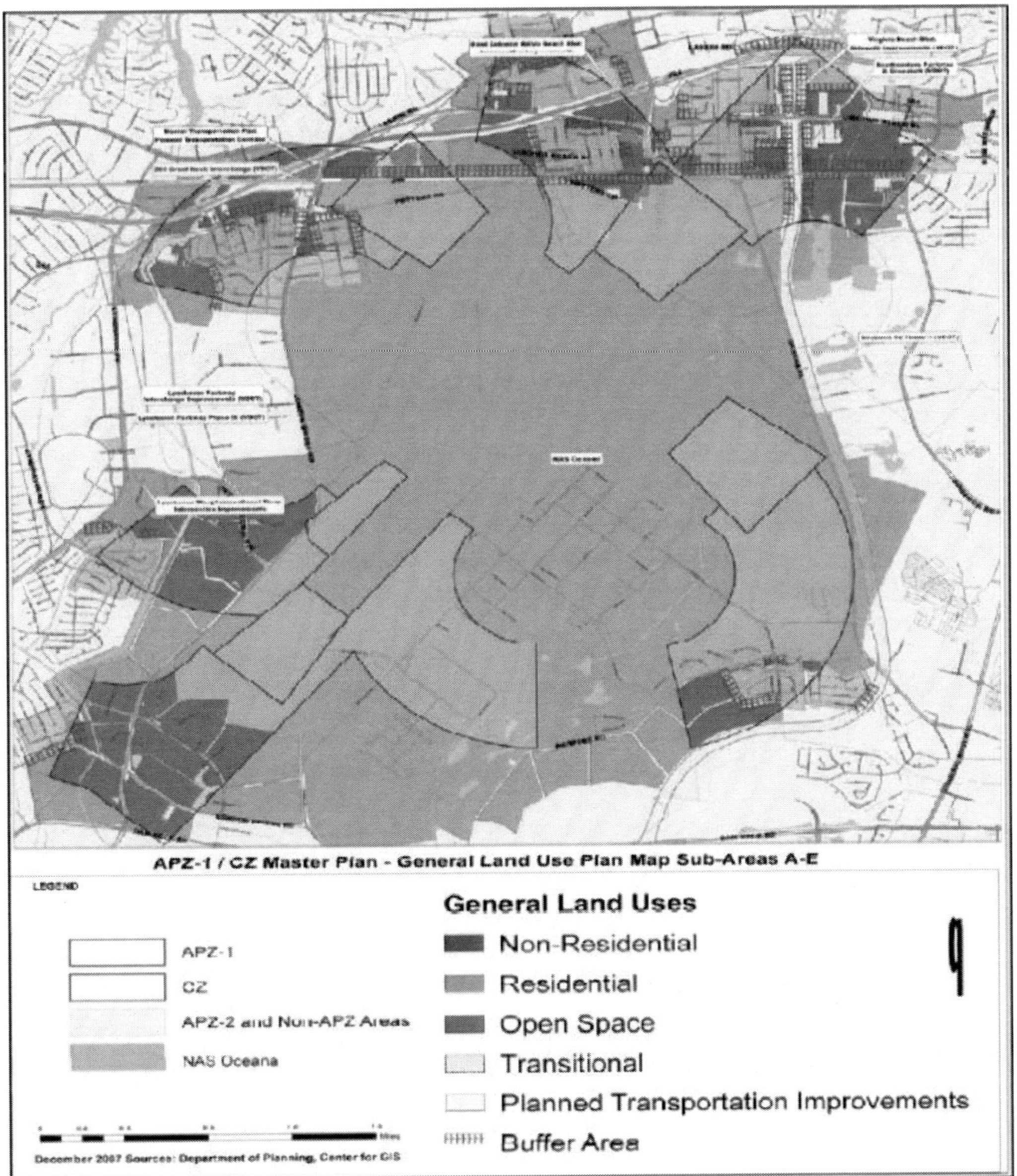

Source: Office of the City Attorney, Virginia Beach, Second Progress Report, N.A.S. Oceana Encroachment, July 1, 2007 – June 30, 2008, Figure 5, p. 19, APZ-1/Clear Zones Master Plan, http://www.vbgov.com/file_source/dept/ mcg/WebPage/ HotTopics/BRAC/Web%20Page/Documents/ nas_oceana_encroachment_2nd_progress_report_2008.pdf.

Notes: APZ is accident potential zone; CZ is clear zone.

Figure 3. GIS Analysis of Naval Air Station Oceana,Virginia Beach,VA (example showing city land use encroachment).

As a result of the GIS analysis, city planners recommended a change to the municipal land ordinance to prevent potential future incompatible development. Naval Air Station Oceana has not been relocated from Virginia Beach.

Mapping Foreclosures

On May 15, 2009, the *New York Times* published an online interactive map showing foreclosures as a percentage of residential properties in each census tract in the New York City region.[17] The map showed census tracts coded by color to represent the foreclosure rate. Moving the cursor over each census tract showed a pop-up window disclosing the foreclosure percentage, the number of foreclosed residences, and the percentage on the non-white population in each tract. In addition, the map allowed the reader to compare foreclosure rates for each year since 2005. *Figures 4* and *5* are snapshots from the map for the years 2005 and 2008.

Using the zoom tool provided with the map allows the reader to zoom in on specific residential properties, represented by red dots, along with street names. This type of visualization, combining detailed geospatial information with demographic and financial data, lends itself to further analysis such as understanding foreclosure patterns and whether proximity to foreclosed properties has an effect on property values. Some researchers call this the "contagion effect" of foreclosed properties.

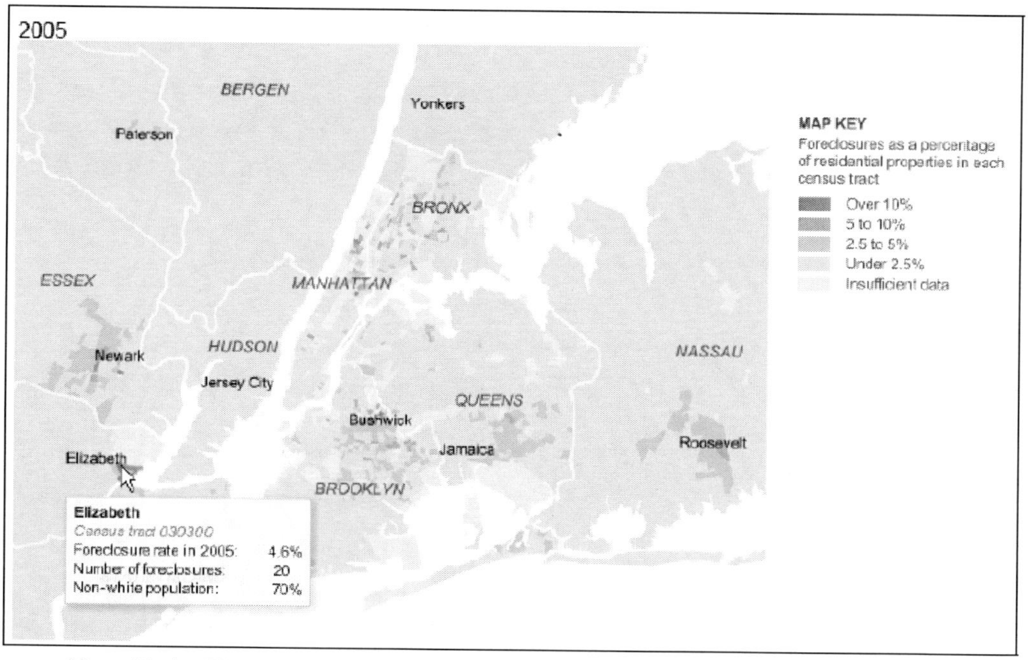

Source: New York Times, May 15, 2009, at http://www.nytimes.com /interactive/ 2009 /05/ 15/nyregion/0515- foreclose.html. Modified by CRS.

Notes: The online interactive version allows the reader to point and click on any census tract in the region. Census tract 030300 is shown here for illustration purposes.

Figure 4. Snapshot of Interactive Map Showing Foreclosure Percentage by Census Tract in the New York City Area, 2005.

One report documented how this effect discounted property values as a function of distance from foreclosed homes, and showed that the discount effect dropped off sharply with distance.[18] This type of spatial analysis of foreclosure effects, with the visualization provided by GIS maps such as the *New York Times* example, can help inform policy makers about the nature and extent of foreclosure patterns, if the underlying data are reliable.

Using the zoom tool provided with the map allows the reader to zoom in on specific residential properties, represented by red dots, along with street names. This type of visualization, combining detailed geospatial information with demographic and financial data, lends itself to further analysis such as understanding foreclosure patterns and whether proximity to foreclosed properties has an effect on property values. Some researchers call this the "contagion effect" of foreclosed properties. One report documented how this effect discounted property values as a function of distance from foreclosed homes, and showed that the discount effect dropped off sharply with distance.[18] This type of spatial analysis of foreclosure effects, with the visualization provided by GIS maps such as the *New York Times* example, can help inform policy makers about the nature and extent of foreclosure patterns, if the underlying data are reliable.

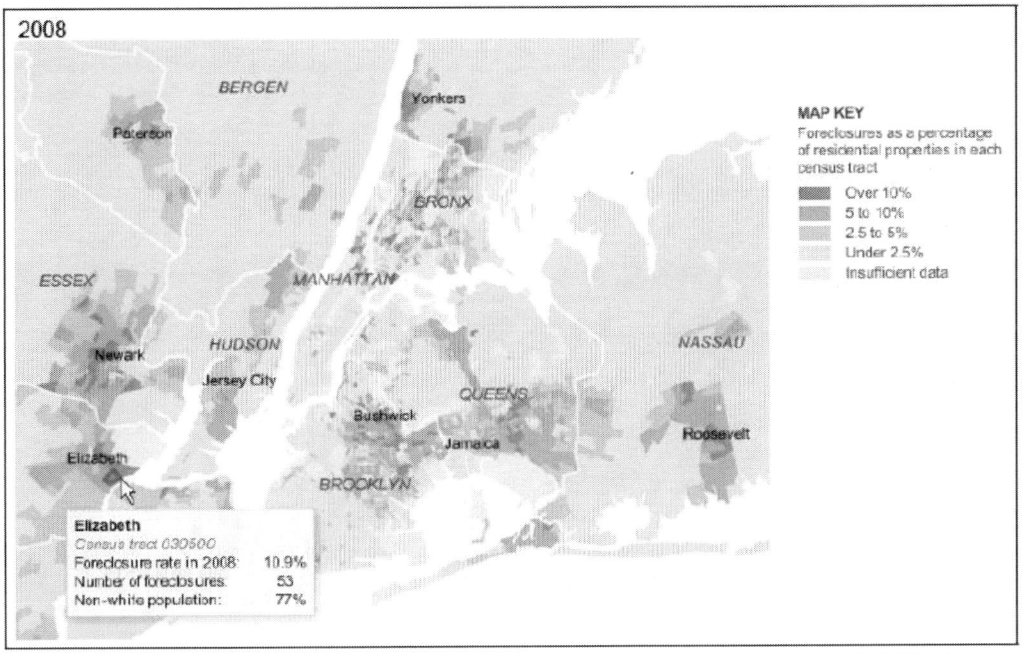

Source: New York Times, May 15, 2009, at http://www.nytimes.com/interactive/ 2009/05/15/nyregion/ 0515- foreclose.html. Modified by CRS.

Notes: The online interactive version allows the reader to point and click on any census tract in the region. Census tract 030500 is shown here for illustration purposes.

Figure 5. Snapshot of Interactive Map Showing Foreclosure Percentage by Census Tract in the New York City Area, 2008.

Social Media and GIS: The March 11, 2011, Japanese Earthquake and Tsunami

Within 36 hours of the March 11, 2011 magnitude 9.0 earthquake off Japan's northeast coast, a small team of GIS experts began a project for a volunteer group called GISCorps. The group intended to gather critical geospatial information to help humanitarian organizations working in areas damaged by the earthquake and tsunami.[19] The project was developed in partnership with Crisis Commons, a social media organization that "seeks to advance and support the use of open data and volunteer technology communities to catalyze innovation in crisis management and global development."[20] According to a source, the project was part of a request from the U.N.

Office for the Coordination of Humanitarian Affairs to provide current information for aid missions in Japan.[21]

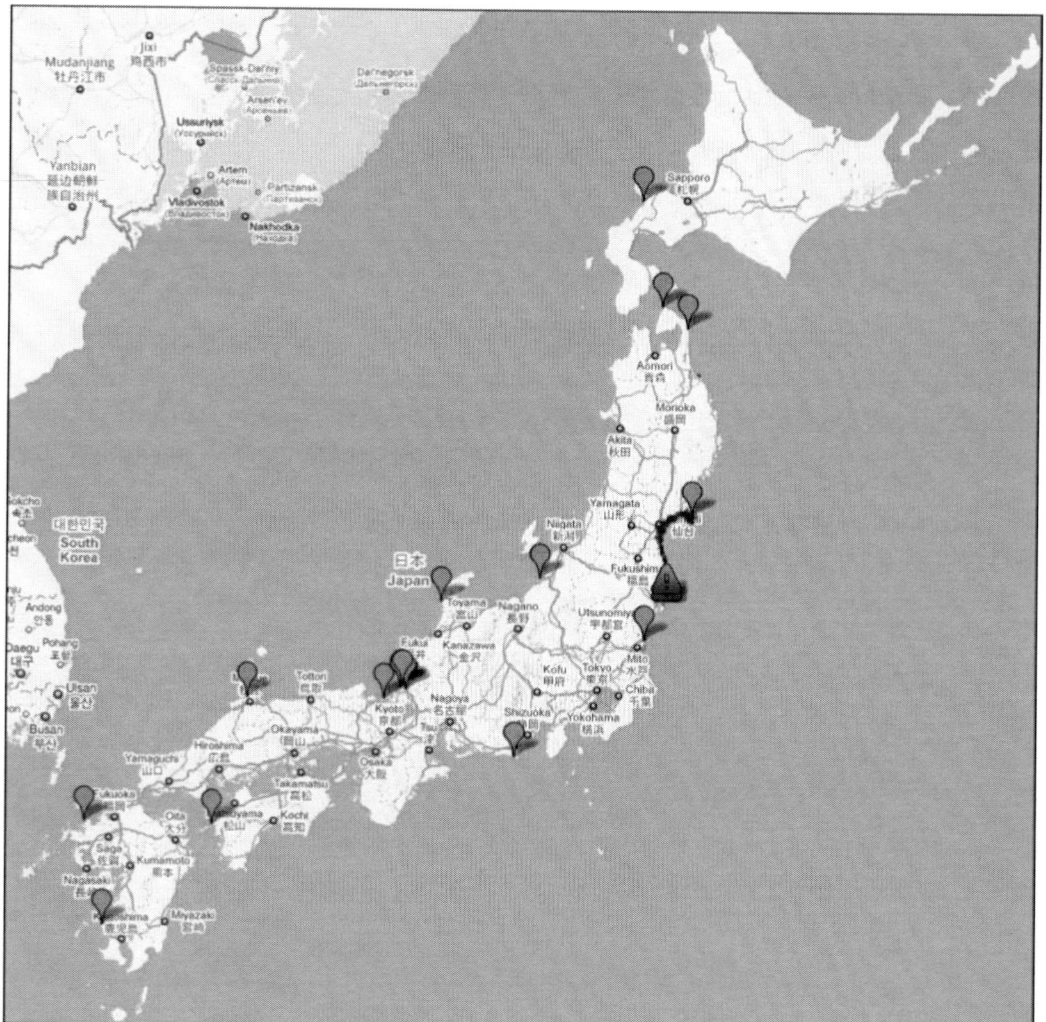

Source: Japan Tohoku Earthquake, CrisisCommons collaboration with GISCorps, http://gis.ats.ucla.edu/japan/.

Notes: Two of seven available data layers have been switched on for this display: nuclear power plants (red balloons); and flood areas (shown as the blue band). The red triangle at the bottom of the blue band indicates the Fukushima I power plant that was damaged by the tsunami.

Figure 6. Example of Interactive Map of Region Affected by the March 11, 2011, Japanese Earthquake and Tsunami.

Within a few days, the team developed an interactive map to display real-time information about evacuation shelters, flooded areas, water distribution centers, cleared roads, and other types of geospatial data of immediate use to citizens and humanitarian workers. Further, several members of the team contacted friends and colleagues in Japan to assist in gathering census records and maps and posting them in GIS layers. This type of information gathering, using on-the-ground sources, is one variant of "crowd-sourcing," or public data collection efforts. In this case, the information was vetted through the team of experts and included as GIS data layers in the interactive map. Crowd-sourcing, also referred to as participatory mapping and sensing, or volunteered geographic information, is becoming increasingly popular as demand for location information increases. As with other examples of participatory data gathering efforts, such as Wikipedia, the accuracy and authoritativeness of the volunteered information is sometimes questionable.

Figure 6 shows the GISCorps/CrisisCommons interactive map with two data layers turned on: the area flooded by the tsunami, and the location of Japan's nuclear power plants, including the damaged Fukushima I plant.

How Geospatial Data Is Managed at the Federal Level

The Federal Geographic Data Committee (FGDC)

OMB Circular A-16 was first issued in 1953 to ensure that federal surveying and mapping activities met the needs of federal and state agencies and the general public and to avoid duplication of effort.[22] In 1967, the circular was revised and included a section on responsibilities for coordination among three federal departments: Department of the Interior, Department of Commerce, and Department of State. In 1990, the circular was revised again and expanded beyond just surveying and mapping to include related geospatial data activities. The 1990 revision also established the Federal Geographic Data Committee (FGDC). The purpose of the FGDC was to promote the coordinated use, sharing, and dissemination of geospatial data nationwide.[23]

Under the revised Circular A-16, 19 members comprise the FGDC. (See *Table 1.)* The U.S. Geological Survey, Department of the Interior, provides administrative support through the FGDC Secretariat.[24] According to Circular A-16, all federal agencies responsible for geospatial data themes (discussed below) are required to be members of the FGDC.

FGDC Coordination Group

In September 2009, the FGDC Steering Committee chartered a Coordination Group, which is charged with identifying, prioritizing, implementing, coordinating, and overseeing strategies and tasks required to support the National Spatial Data Infrastructure (NSDI, discussed in the next section).[25] One of its objectives is to advance intra- and inter-government coordination to ensure implementation of NSDI and to enhance the management of the federal geospatial enterprise and national geospatial data sets. The other objectives appear to align with the goal of improving coordination within the federal government, to meet the responsibilities outlined in Circular A-16. The members of the Coordination Group include agency and department members of the FGDC (Table 1), and 11 additional members (two non-voting) representing agencies of the federal government. Individual members of the Coordination Group are responsible for coordinating their agencies' Circular A-16 responsibilities, according the Coordination Group charter.

Table 1. Members of the Federal Geographic Data Committee (FGDC)

Dept. of Agriculture	Environmental Protection Agency
Dept. of Commerce	Federal Emergency Management Agency
Dept. of Defense	General Services Administration
Dept. of Energy	Library of Congress
Dept. of Health and Human Services	National Aeronautics and Space Administration
Dept. of Housing and Urban Development	National Archives and Records Administration
Dept. of the Interior (Chair)	National Science Foundation
Dept. of Justice	Tennessee Valley Authority
Dept. of State	
Dept. of Transportation	Office of Management and Budget (Co-Chair)

The National Spatial Data Infrastructure (NSDI)

OMB Circular A-16 also called for development of a national digital spatial information resource to enable the sharing and transfer of spatial data between producers and users, linked by criteria and standards. This national digital spatial information resource became what is now known as the National Spatial Data Infrastructure (NSDI).[26] In an August 19, 2002, revision, Circular A-16 affirmed the NSDI as "the technology, policies, standards, human resources, and related activities necessary to acquire, process, distribute, use, maintain, and preserve spatial data."

Circular A-16 directs the FGDC to lead and support the NSDI strategy, spatial data policy development, management, and operational decision making. As the overall coordinating entity for the NSDI, the FGDC has broad responsibilities that include all spatial data and GIS activities financed directly or indirectly, in part or in whole, by federal funds.

The FGDC facilitates the NSDI in cooperation with organizations from state, local, and tribal governments, the academic community, and the private sector. As specified in Circular A-16, cooperation is necessary to realize the overall vision of the NSDI: to assure that spatial data from multiple sources—not just federal sources—are available and easily integrated to enhance the understanding of our physical and cultural world. There are five components[27] of the NSDI: (1) data themes; (2) metadata; (3) National Spatial Data Clearinghouse; (4) standards; and (5) partnerships. (Refer to the box above for definitions of common geospatial and GIS terms.)

Data Themes

Data themes are electronic records and coordinates for a topic or subject, such as elevation or vegetation. Themes that are the core group of the most commonly used set of base data are known as framework data, which are (1) geodetic control, (2) orthoimagery, (3) elevation and bathymetry, (4) transportation, (5) hydrography, (6) cadastre, and (7) governmental units.

Metadata

Metadata are information about the data, its content, source, accuracy, method of collection, and other descriptions that help ensure the data are used appropriately. They help ensure that the data are authoritative, and that any resulting analysis is credible. OMB Circular A-16 specifies that all spatial data collected or derived directly or indirectly using federal funds will have FGDC metadata.

National Spatial Data Clearinghouse

The Clearinghouse is an electronic service providing access to documented spatial data and metadata from distributed data sources. The Clearinghouse is intended to provide access to NSDI for spatial data users. According to Circular A-16, all spatial data collected by federal agencies will be made available through the Clearinghouse.

Standards

These are common and repeated rules, conditions, guidelines or characteristics for data, and related processes, technology, and organization. OMB Circular A-16 specifies that international standards and protocols must be used for NSDI, to broaden the global use of federal data and services. The FGDC is responsible for developing and promulgating the standards after receiving broad input from data users and providers. Also, Circular A-16 directs FGDC to adopt national and international standards in lieu of federal standards whenever possible.

Partnerships

OMB Circular A-16 directs federal agencies to promote and use partnerships that promote cost-effective data collection, documentation, maintenance, distribution, and preservation strategies that leverage the federal resources. In addition to federal, state, and tribal governments, these partnerships are to include private-sector geographic, statistical, demographic, and other business information providers and users.

Most recently, OMB issued supplementary guidance to Circular A-16 on November 10, 2010.[28] The supplemental guidance labels geospatial data as a capital asset, and refers to its acquisition and management in terms analogous to financial assets to be managed as a National Geospatial Data Asset Portfolio. (See CRS Report Rxxxxx for a discussion of the most recent supplemental guidance to Circular A-16.)

Other Activities and Components of FGDC and NSDI

The National Map

Introduced in 2001, the National Map is envisioned as a consistent framework for geographic knowledge nationwide, and will be available as an online and interactive map service, according to the USGS.[29] The National Map is the future product—the next generation—that would supplant the paper versions of topographic maps that the USGS has produced for decades. Topographic maps are probably the USGS product most familiar to the majority of Americans. The USGS plans to replace all of the current topographic maps for the conterminous 48 states over the next three years with a new map product: “US Topo.[30]

The National Map would allow users to combine geographic information from other sources with the USGS topographic foundation data. The National Map would provide information such as high-resolution digital imagery from satellites and aerial photographs, high-resolution surface elevation data, land cover data, geographic names, and other features. Currently, the National Map is in an initial stage that can provide nationwide coverage at limited resolutions for transportation, hydrography, elevation, land cover, and cultural features. According to the USGS, the National Map will capture and integrate data in a process of continuous update, rather than by regularly scheduled cycles of review and revision. The National Map will face challenges, however, in integrating data from a variety of sources, perhaps at different scales and different resolutions, and in managing inconsistent or incomplete metadata.

Geospatial One-Stop

The Geospatial One-Stop portal is the official means of accessing metadata resources, published through the National Spatial Data Clearinghouse and managed in NSDI. [31] Geospatial One-Stop focuses on the discovery and access of geospatial information.[32] The Geospatial One-Stop is described as one of the three national geospatial initiatives that share the goal of building the NSDI along with FGDC itself and the National Map. These three components have different foci: the FGDC focuses on policy, standards, and advocacy; and the National Map focuses on integrated, certified, base mapping content. In FY2010, according to the USGS, Geospatial One-Stop focused on integration with the Obama Administration's Open Government initiative through Data.gov.[33] (Data.gov is discussed further below.)

USGS Geospatial Liaison Network

The USGS Geospatial Liaison Network consists of USGS employees who serve as liaisons in NSDI partnership offices across the country.[34] The liaisons are intended to represent and coordinate the National Geospatial Program (NGP)[35] initiatives in state and local agencies, in addition to other federal agencies, in support of NSDI, The National Map, and Geospatial One-Stop. The liaisons work with statewide coordinating councils and seek partnerships with not-forprofit organizations, the private sector, universities, and consortia to support the goals of NSDI. According to the USGS, each liaison is the "local face" of the USGS, NSDI, and NGP.

Each state is assigned a liaison under the network, although some liaisons may cover more than one state. The liaisons commonly work with the formal state GIS coordinators or councils, or with other individuals or regional groups in states where a formal GIS or geospatial coordinator or council does not exist. A large portion of the liaison's efforts is devoted to coordinating with state-level and other stakeholders on geospatial data acquisition and data maintenance. In states with large federal land holdings, such as some western states, geospatial liaisons may devote relatively more time to coordinating with federal land management agencies such as the Forest Service, National Park Service, Fish and Wildlife Service, or Bureau of Land Management.[36]

Data.gov

Data.gov was launched in 2009 under the Obama Administration with the primary goal of improving access to federal data generated by the Executive Branch.[37] During its first year,

Data.gov expanded its accessible raw data to over 270,000 records, 90% of which were geospatial data.[38] Geospatial records in Data.gov are shared with Geospatial One-Stop. Although Data.gov is not exclusively an activity under FGDC and NSDI, its prominence under the Obama Administration's Open Government initiative together with its preponderance of geospatial data make it a growing location for access to and coordination with geospatial elements managed under FGDC.

APPENDIX. LIST OF ACRONYMS

ARRA	American Recovery and Reinvestment Act of 2009
APZ	Accident Potential Zone
BLM	Bureau of Land Management
BRAC	Base Realignment and Closure
COGO	Coalition of Geospatial Organizations
CZ	Clear Zone
DFIRMs	Digital Flood Insurance Rate Maps
FEMA	Federal Emergency Management Agency
FGDC	Federal Geographic Data Committee
FIRMs	Flood Insurance Rate Maps
GAO	Government Accountability Office
GIS	Geographic Information Systems
GPS	Global Positioning System
IFTN	Imagery for the Nation
LIDAR	Light Detection and Ranging
MAPPS	Management Association for Private Photogrammetric Surveyors
NGAC	National Geospatial Advisory Committee
NGP	National Geospatial Program
NILS	National Integrated Land System
NRC	National Research Council
NSDI	National Spatial Data Infrastructure
NSGIC	National States Geographic Information Council
OMB	Office of Management and Budget
USGS	U.S. Geological Survey
USFS	U.S. Forest Service
VGI	Volunteered Geographic Information

End Notes

1 The development and commercial availability of Global Positioning System (GPS) data and the integration of these data with digital maps has led to the popular handheld or dashboard navigation devices used daily by millions.

2 The National Geospatial Advisory Committee, The Changing Geospatial Landscape, January 2009, p. 10, http://www.fgdc.gov/ngac/NGAC%20Report%20-%20The%20Changing%20 Geospatial%20Landscape.pdf. Hereinafter referred to as NGAC, The Changing Geospatial Landscape, January 2009.

[3] The Obama Administration issued supplemental guidance to OMB Circular A-16 on November 10, 2010, which labeled federal geospatial data a capital asset, and referred to its acquisition and management in terms analogous to financial assets. See Vivek Kundra, Federal Chief Information Officer, Geospatial Line of Business OMB Circular A16 Supplemental Guidance, Office of Management and Budget, November 10, 2010, http://www. white house.gov/sites/ default/files/omb/memoranda/2011/m11-03.pdf.

[4] U.S. Geological Survey, Geographic Information Systems, http://egsc.usgs.gov /isb/pubs/ gis_poster/#what.

[5] National Research Council, Successful Response Starts With a Map: Improving Geospatial Support for Emergency Management, Washington, DC, 2007, p. 15.

[6] New York State Department of Environmental Conservation, Center for Technology in Government, Sharing the Costs, Sharing the Benefits: the NYS GIS Cooperative Project, Project Report 95-4, Albany, NY, 1995, p. 7, http://www.ctg.albany.edu/publications /reports/sharing_the_costs/sharing_the_costs.pdf.

[7] For example, thousands of amateur geospatial enthusiasts are forming mapping parties, using personal navigation devices to create their own street maps. See http://www. OpenStreetMap.org. Information derived from such groups is referred to as volunteered geographic information (VGI).

[8] U.S. General Accounting Office (now the Government Accountability Office), Geospatial Information: Better Coordination Needed to Identify and Reduce Duplicative Investments, GAO-04-703, June 23, 2004, p. 13. Hereinafter referred to as GAO (2004).

[9] GAO (2004).

[10] National Research Council, Successful Response Starts With a Map, 2007, p. 3.

[11] Section 575 of P.L. 103-325 requires the Director of FEMA to assess the need to revise and update all floodplain areas and flood risk zones identified.

[12] For more information on the flood map modernization initiative and flood insurance, see CRS Report R41056, Mandatory Flood Insurance Purchase in Remapped Residual Risk Areas Behind Levees, by Rawle O. King.

[13] Most states do not have comprehensive LIDAR coverage. North Carolina is an exception, having nearly complete coverage because it implemented a statewide LIDAR program, in part to improve the accuracy of state flood plain maps in the wake of Hurricane Floyd in 1999. See http://www.ncfloodmaps.com/pubdocs/FAQs.pdf.

[14] See MAPPS, at http://www.mapps.org/.

[15] For a list of the COGO member organizations, see http://www.urisa.org/cogo.

[16] Peter Fimrite, "Wildfires Raging Through Southern California," San Francisco Chronicle, November 16, 2008, online article via SFGate, http://www.sfgate.com/cgi-bin/article.cgi?f= /c/a/2008/11/15/MNC7145F2D.DTL.

[17] Mathew Bloch and Janet Roberts, "Mapping Foreclosures in the New York Region," New York Times, May 15, 2009, at http://www.nytimes.com/interactive/ 2009/05/15/nyregion /0515-foreclose.html.

[18] John P. Harding, Eric Rosenblatt, and Vincent W. Yao, "The Contagion Effect of Foreclosed Properties," Social Science Research Network Working Paper, July 15, 2008.

[19] WAMU 88.5 American University Radio, broadcast of April 21, 2011, http://wamu. org/news/11/04/21/ university_of_maryland_professor_helps_digitize_disaster_relief.php.

[20] CrisisCommons, http://crisiscommons.org/about/.

[21] Japan Tohoku Earthquake GISCorps/CrisisCommons collaboration, http://gis.ats.ucla.edu/ japan/.

[22] OMB Circular A-16, Appendix C.

[23] GAO (2004), p. 11.

[24] See the USGS National Geospatial Program, at http://www.usgs.gov/ngpo/index.html.

[25] See FGDC Coordination Group Charter, at http://www.fgdc.gov/participation/coordination-group/coordination-group-charter.

[26] On April 11, 1994, President Clinton issued Executive Order 12906, which specified that the FGDC shall coordinate development of the NSDI.

[27] These descriptions are drawn from Circular A-16 Revised, August 19, 2002, at http://www.whitehouse.gov/omb/ circulars_a016_rev/.

[28] Office of Management and Budget, Geospatial Line of Business OMB Circular A-16 Supplemental Guidance, November 10, 2010.

[29] See USGS, The National Map, at http://nationalmap.gov/index.html.

[30] USGS Budget Justification for FY2011, p. I-38.

[31] USGS. Citied in the FGDC 2009 Annual Report, p. 31, http://www. fgdc.gov /library/ whitepapers-reports/ annual%20reports/2009/2009-AR.pdf.

[32] The website is called geodata.gov and is available at http://gos2.geodata.gov/wps/portal/gos.

[33] FGDC 2010 Annual report, at http://www.fgdc.gov/library/whitepapers-reports/annual% 20reports/2010/web-version/ index_html.

[34] A description of the program, and a link to a list of the liaisons, is provided at http://www.usgs.gov/ngpo/ngp_liaisons.html#. Additional information about the broader geospatial partnerships is provided at http://liaisons.usgs.gov/geospatial/.

[35] The USGS National Geospatial Program (NGP) organizes, maintains, and publishes the geospatial baseline of the nation's topography, natural landscape, and built environment, including transportation features. The National Map is a subcomponent of the NGP within the USGS organizational structure. See the USGS Budget Justification, pp. I-29 to I48 for a fuller description of NGP.

[36] Ibid.

[37] See Data.gov at http://www.data.gov/about.

[38] FGDC 2010 Annual report.

In: Federal Geospatial Information
Editors: Andrew Halpern and Mark Troupp ISBN 978-1-62257-572-5

Chapter 2

ISSUES AND CHALLENGES FOR FEDERAL GEOSPATIAL INFORMATION*

Peter Folger

SUMMARY

Congress has recognized the challenge of coordinating and sharing geospatial data from the local, county, and state level to the national level, and vice versa. The cost to the federal government of gathering and coordinating geospatial information has also been an ongoing concern. As much as 80% of government information has a geospatial component, according to various sources.

The federal government's role has changed from being a primary provider of authoritative geospatial information to coordinating and managing geospatial data and facilitating partnerships. Congress explored issues of cost, duplication of effort, and coordination of geospatial information in hearings during the 108th Congress. However, challenges to coordinating how geospatial data are acquired and used—collecting duplicative data sets, for example—at the local, state, and federal levels, in collaboration with the private sector, are not yet resolved. Two bills introduced in the 112th Congress, H.R. 1620 and H.R. 4322, would address aspects of duplication and coordination of geospatial information.

The federal government has recognized the need to organize and coordinate the collection and management of geospatial data since at least 1990, when the Office of Management and Budget (OMB) revised Circular A-16 to establish the Federal Geographic Data Committee (FGDC) and to promote the coordinated use, sharing, and dissemination of geospatial data nationwide. OMB Circular A-16 also called for development of a national digital spatial information resource to enable the sharing and transfer of spatial data between users and producers, linked by criteria and standards. Executive Order 12906, issued in 1994, strengthened and enhanced Circular A-16, and specified that FGDC shall coordinate development of the National Spatial Data Infrastructure (NSDI). On November 10, 2010, OMB issued supplemental guidance to Circular A-16 that labels geospatial data as a "capital asset," and refers to its acquisition and management in terms analogous to financial assets to be managed as a National

---

* This is an edited, reformatted and augmented version of a Congressional Research Service publication, CRS Report *for Congress* R41826, *from* www.crs.gov, prepared for Members and Committees of Congress, *dated* April 27, 2012.

Geospatial Data Asset Portfolio. It will likely take some time, and several budget cycles, to track whether agencies are adhering to the "portfolio-centric model" of geospatial data management outlined in the supplemental guidance. The 112th Congress may examine its oversight role in the implementation of OMB Circular A-16, particularly in how federal agencies are coordinating their programs that have geospatial assets.

The high-level leadership and broad membership of the FGDC—10 cabinet-level departments and 9 other federal agencies—highlights the importance of geospatial information to the federal government. Questions remain, however, about how effectively the FGDC is fulfilling its mission. Has this organizational structure worked? Can the federal government account for the costs of acquiring, coordinating, and managing geospatial information? How well is the federal government coordinating with the state and local entities that have an increasing stake in geospatial information? What is the role of the private sector?

State-level geospatial entities, through the National State Geographic Information Council, also embrace the need for better coordination. However, the states are sensitive to possible federal encroachment on their prerogatives to customize NSDI to meet the needs of the states. In early 2009, several proposals were released calling for efforts to create a national Geospatial Information System (GIS). Language in the proposals attempted to make the case for considering such efforts part of the national investment in critical infrastructure. Congress may consider how a national GIS or geospatial infrastructure would be conceived, perhaps drawing on proposals for these national efforts and how they would be similar to or differ from current efforts.

INTRODUCTION

Historically, the federal government has been a primary provider of authoritative geospatial information, but some argue that consumer demand for spatial information has triggered a major shift toward local government and commercial providers.[1] The federal government has shifted, with some important exceptions, to consuming rather than providing geospatial information from a variety of sources. As a result, the federal government's role has shifted as well toward coordinating and managing geospatial data and facilitating partnerships among the producers and consumers of geospatial information in government, the private sector, and academia. The challenges to coordinating how geospatial data are acquired and used—collecting duplicative data sets, for example—at the local, state, and federal levels, in collaboration with the private sector, are long-standing and not yet resolved.

In 2003 and 2004 the Subcommittee on Technology, Information Policy, Intergovernmental Relations, and the Census, part of the House Committee on Government Reform, held two hearings on the nation's geospatial information infrastructure. A common theme to both hearings was the challenge of coordinating and sharing geospatial data between the local, county, state, and national levels. Quantifying the cost of geospatial information to the federal government has also been an ongoing concern for Congress. At the hearing in 2003, Representative Adam Putnam stated:

> We need to understand what programs exist across the government, how much we're spending on those programs, where we're spending that money, how efficiently, or perhaps inefficiently, we share data across Federal agency boundaries, how we separate security-sensitive geospatial data from those open for public use, and how we efficiently, or perhaps inefficiently, coordinate with State and local governments and tribes.[2]

The explosion of geospatial data acquired at the local and state levels, for their own purposes and in conjunction with the private sector, underscores the long-recognized need for better coordination between the federal government and local and state authorities. At the same time, coordinating, managing, and facilitating the production and use of geospatial information from different sources, of different quality, and which was collected with specific objectives in mind has been a challenge. The federal government has recognized this challenge since at least 1990, when the Office of Management and Budget (OMB) revised Circular A-16 to establish the Federal Geographic Data Committee (FGDC) and to promote the coordinated use, sharing, and dissemination of geospatial data nationwide.[3]

Executive Order 12906, issued in 1994, was intended to "strengthen and enhance" the general policies in Circular A-16. [4] Executive Order 12906 specified that the FGDC shall coordinate the federal government's development of the National Spatial Data Infrastructure (NSDI).[5]

Further, EO12906 called for the establishment of a National Geospatial Data Clearinghouse to address data standardization, make geospatial data publically available, and address redundancy and incompatibility of geospatial information. Circular A-16 was itself revised in 2002, adding the Deputy Director of Management at OMB as the vice-chair of the FGDC to serve with the Secretary of the Interior, who chairs the committee.

The high-level leadership and broad membership of the FGDC—10 cabinet-level departments and 9 other federal agencies—highlights the importance of geospatial information to the federal government.[6] In fact, supplemental guidance to Circular A-16, issued by Federal Chief Information Officer Vivek Kundra on November 10, 2010, referred to federal geospatial data as a *capital asset.*[7] Questions remain, however, about how effectively the FGDC is fulfilling its mission. Has this organizational structure worked? Can the federal government account for the costs of acquiring, coordinating, and managing geospatial information? How well is the federal government coordinating with state and local entities that increasingly rely on geospatial information? What is the role of the private sector? In its oversight capacity, the 112th Congress may consider these questions from the viewpoint of reducing duplication and costs to the federal government.

The chapter discusses issues that may be of interest to Congress—managing, sharing, and coordinating geospatial information—and includes examples of legislation. The chapter also summarizes a diverse set of recommendations and proposals from different non-governmental organizations for how to improve the coordination and management of geospatial information at the federal and state levels.

Issues with Organization and Management, Data Sharing, and Coordination

Producing floodplain maps, conducting the Census, planning ecosystem restoration, and assessing vulnerability and responding to natural hazards such as hurricanes, earthquakes, and tsunamis are examples of how federal agencies use GIS and geospatial information to meet national needs.

Some view federal government data as inherently geospatial. According to the Department of the Interior, the amount of government information that has a geospatial component—such as address or other reference to a physical location—is as much as 80%.[8] Given the ubiquity of geospatial information throughout the federal government, and despite the long history of efforts to manage and coordinate such data articulated in OMB Circular A-16 and its antecedents (See *Appendix* for a history of Circular A-16 and its federal policy lineage), ongoing challenges to handling federal geospatial information can generally be divided into three overarching questions:

- What is the best way to organize and manage the vast array of geospatial information that is acquired at many levels and that has a variety of potential uses?
- What is the best way to share data, particularly among local, state, and federal stakeholders, each of whom may have a need for the same or similar data?
- What is the best way to coordinate among federal agencies, such as the administration and management by different agencies of all the federal lands in the United States?

Organization and Management of Geospatial Data

It could be argued that some level of duplication of effort, and of inefficiency in the management and sharing of geospatial information, will always exist across a vast federal bureaucracy in which a majority of government information has some geospatial component. It could also be argued that the size of the federal bureaucracy is only one contributing factor. Surveying and mapping activities themselves are prone to duplication of effort among the different missions and goals of the executive branch.[9]

The need to organize and manage geospatial data among federal agencies and among the federal government, local and state authorities, the private sector, and academia is a recurring theme. It recurs, in part, because it is widely recognized that collecting data multiple times for the same purpose is wasteful and inefficient, yet it continues to occur. Alternatively, geospatial data could be collected once to meet the requirements of several users. For example, geospatial data gathered by a local government could be made useful to the state or federal government if the data meet a set of basic and consistent guidelines and protocols. Organizational structures exist at the federal and state levels to identify and promulgate the efficient sharing, transfer, and use of geospatial information, but arguably they have not fully achieved the goal of seamlessly coordinating disparate types of geospatial data. Ideally, these efforts would produce a national spatial data infrastructure, or NSDI, and that appears to have been the intent of EO12906 and the subsequent revisions to Circular A-16 in 2002.

In one sense, the FGDC exists to foster development and implementation of the NSDI.[10] The NSDI includes the processes and relationships that facilitate data sharing across all levels of government, academia, and the private sector. Ultimately, it is intended to be the base resource and structure among geospatial data providers and users at the national, state, and local level. Yet, some members of the geospatial community have indicated that the past efforts to create a national spatial data infrastructure have not met expectations, and have called for a new effort to build a "national GIS" or a "NSDI 2.0." (See section below: "A National GIS?") In addition to promoting the efficiency and interoperability of such a national system, some promote NSDI as "digital infrastructure" on par with other parts of the nation's critical infrastructure—such as roads, pipelines, telecommunications—and underscore its role in the national economy and in national security.

When Circular A-16 was issued in 1953, it aimed to avoid duplication of effort, and included details about coordinating federal mapping activities.[11] As digital geospatial data became more widespread, revisions to Circular A-16 in 1990 and 2002 extended coordination of federal efforts to include digital data, and broadened the mandate to coordinate federal geospatial activities. These efforts to improve management and coordination continue: On March 3, 2006, OMB issued a memorandum asking selected departments and agencies to each designate a senior agency official to take authority and responsibility for geospatial information issues.

The memorandum emphasized that "... through further coordination, we will maximize our buying and maintaining of geospatial investments instead of independently investing in potentially duplicative and costly data and capabilities."[12]

The March 3, 2006, memorandum identified 15 departments and 12 independent agencies that should designate an official at the Assistant Secretary or equivalent level. Most recently, OMB issued supplementary guidance to Circular A16 on November 10, 2010 (discussed in the next section).

Supplemental Guidance to OMB Circular A-16, November 10, 2010

Geospatial Data as a Capital Asset

On November 10, 2010, the Obama Administration issued a memorandum providing supplemental guidance to the implementation of OMB Circular A-16.[13] The supplemental guidance labels geospatial data as a capital asset, and refers to its acquisition and management in terms analogous to financial assets. Specifically, it refers to geospatial information as part of a National Geospatial Data Asset (NGDA) Portfolio.

The supplemental guidance states that federal investments in geospatial data "...were largely uncoordinated and often lacked transparency, and sometimes resulted in data deficiencies, lack of standardization, inefficient use of resources, lack of interoperability, or inability to share data."[14] To address those issues, the supplemental guidance sets forth its goal of a portfolio-centric model that "...cures the single agency, stovepipe model by applying consistent policy, improved organization, better governance, and understanding of the electorate to deliver outstanding results."[15] The supplemental guidance appears to echo the same concerns regarding management, coordination, and sharing previously identified. However, it casts the solution to these challenges in terms of managing a portfolio of investments, described as "...the process of tracking, maintaining, expanding, and aligning assets to address and solve the business needs of an enterprise."[16]

The document notes that a key goal of the portfolio management approach is to enable the FGDC, through its Steering Committee, to make "informed decisions" about short- and long-term priorities for NGDA themes and datasets as well as for "collaboration targets" across agencies for dataset development and funding. The approach to NGDA portfolio management would consist of "[t]he inventory, selection, organization, management, evaluation, monitoring, and setting of Federal geospatial dataset priorities to ensure that NGDA Datasets are available to support the mission needs of the Federal Government and its partners, as determined by Federal agencies and their partners and as recommended to OMB."[17]

It is not clear yet whether this "new" approach to organizing and managing federal geospatial information is significantly different from the current approach. Perhaps a financial asset portfolio management approach could provide a more internally consistent way of identifying, classifying, and managing diverse geospatial assets across multiple departments and agencies. Whether this approach will be put into practice by each individual department and agency, and how consistently the "capital asset" approach will be applied across the federal government, remains to be seen. In the executive branch, OMB is the most likely entity capable of assessing the government-wide implementation of the geospatial-information-as-financial-assets approach. Congress in its oversight capacity could choose to ask whether the approach has been implemented and, if so, whether it produced the desired results.

Managing Geospatial Assets Within the Annual Budget Cycle

The supplemental guidance also lays out a process for managing geospatial assets within the annual budget cycle, calling it an annual investment review process. This process could give the agencies with geospatial assets a potentially more visible role in obtaining funding to acquire and manage geospatial data. The supplemental guidance notes that this process could increase the geospatial community's effectiveness by addressing a "disconnect" between agency Chief Financial Officers and managers responsible for an agency's geospatial assets:

> The players traditionally active in the Federal agency budget formation process, most notably the agency CFO community, rarely have expertise in geospatial management or issues. At the same time, those with significant geospatial expertise rarely have a distinct role in the budget process. The fact that so much Federal geospatial spending is subsumed unidentifiably within otherprogram budgets, and therefore opaque to the CFO community, is one reason for the disconnect.[18]

Whether and how the new guidance will affect FGDC management will likely be borne out in the budget priorities for agencies in the annual budget requests, the level of funding for those priorities through the annual appropriations cycle, and in program implementation. The 112th Congress might compare and contrast the broad portfolio management approach outlined in the supplemental guidance against funding requests by agencies that invest in and manage geospatial information.

Data Sharing Between Local, State, and National Levels

The National Research Council (NRC) has reported that the value of geospatial data is better understood at the county level than it was in the past, especially land parcel, or cadastral, data.[19] The benefits of sharing geospatial data so that what is produced locally can be used for national needs, however, are not as widely acknowledged.[20] In the case of land parcel data specifically, many local governments create data for their own use and do not see how a national effort would bring local benefits. The NRC notes, however, that the need for complete national land parcel data has become urgent particularly for at least one application—emergency response. During the Hurricane Katrina disaster, some critical land parcel data that was needed by emergency responders, public officials, and even insurance companies was not readily available or did not exist.[21] Further, the NRC report asserts that many of the property fraud cases associated with the hurricanes of 2005 were the direct result of poor or nonexistent geospatial data, specifically land parcel data.[22]

Challenges to Coordinating

Several efforts to coordinate geospatial data among federal agencies have proven difficult to achieve. The National Map is an example of a work-in-progress attempting to integrate data from a variety of sources and produce a product that is widely available and useful to many users.[23] In an example cited by the GAO in its 2003 testimony, the U.S. Forest Service (USFS) tried to create a national-level GIS for the forest ecosystem, but had to reconcile data from a variety of incompatible locally developed systems, which used a variety of standards for each forest and district. Most of the USFS effort went into reconciling the different data sets. Ultimately the USFS had to adopt the lowest-resolution format to maintain full coverage of all the forests, and could not use the higher-resolution local data.[24]

The National Integrated Land System (NILS) is another example of an ongoing effort to coordinate and integrate disparate federal land data among several agencies. The federal government owns approximately 650 million acres, about 29% of all land in the United States. The Bureau of Land Management (BLM) is the designated custodian for federal land parcel information and ownership status.[25] Three federal agencies in addition to the BLM administer most federal lands: the USFS, Fish and Wildlife Service, and the National Park Service.[26] In an effort to develop a single representation of federal lands, the BLM and USFS launched NILS, billed as a partnership among the federal agencies and states, counties, and private industry, to provide a single solution to managing federal land parcel information in a GIS environment.

A limited amount of federal land data is available through NILS, which is currently in a project or prototype phase, and the project makes current information and tools available through its GeoCommunicator component.[27]

Floodplain Mapping:
An Example of Data Sharing

In one sense, floodplain mapping represents an archetypical example of how GIS and geospatial data can be shared to fulfill national and local needs. In this case, the shared need is for accurate floodplain maps. Floodplain mapping also exemplifies the challenge of collecting and sharing geospatial data of sufficient accuracy to meet the needs of local, state, and federal data users and authorities. A 2009 report issued by the National Research Council (NRC) observed that high-quality, digital mapping of floodplains using the most accurate elevation data is essential to communicate flood hazards, set flood insurance rates, and regulate development in flood-prone areas. Between 2003 and 2008, the Federal Emergency Management Agency (FEMA) invested approximately $1 billion in the Map Modernization Program, a large-scale effort to collect new elevation data, update existing data, and digitize older paper flood maps. State governments and local partners also contributed considerable funding to the effort.

The FEMA effort produced digital flood maps covering 92% of the nation's population; however, according to the NRC report only 21% of the population has flood maps that fully meet FEMA's own data quality standards. As a result, insurance companies, lenders, realtors, and property owners who depend on the flood maps to determine flood insurance needs, plan for development, and prepare for floods still have to deal with uncertainties inherent in the less accurate flood maps. A fundamental requirement for accurate flood maps is accurate elevation data, which are used to draw the boundaries for the 1-in-100 chance annual flood hazard (sometimes referred to as the 100-year flood). Part of the challenge of producing high quality digital floodplain maps is that there is no single nationwide set of elevation data of sufficient resolution and accuracy to make floodplain maps that meet FEMA requirements. The USGS National Elevation Dataset is a primary data source that FEMA uses to produce flood maps, but it has a level of uncertainty about 10 times larger than FEMA defines as acceptable for floodplain mapping, according to NRC. The USGS National Elevation Dataset does include some high-resolution, more accurate elevation data, but most of the USGS dataset is of poorer resolution. Alternate sources of more accurate elevation data exist, but are not available nationwide.

One of these sources is provided using LIDAR (an acronym for Light Detection and Ranging), which can be used to collect high-resolution elevation data. For example, North Carolina instigated a state-wide LIDAR program, in part, to improve the accuracy of floodplain maps in the wake of hurricane Floyd in 1999. As a result, the state has nearly complete LIDAR coverage. And, floodplain maps for nearly all the counties in North Carolina meet or exceed national flood hazard data quality thresholds. Because of this data gap at the national level, the 2009 NRC report recommends that FEMA should increase its collaboration with federal, state, and local government agencies to acquire high-resolution and accurate elevation data across the nation.

Sources: National Research Council, Mapping the Zone: Improving Flood Map Accuracy, Washington, DC, 2009, Summary; Recommendations; p. 38; and CRS Report R41056, Mandatory Flood Insurance Purchase in Remapped Residual Risk Areas Behind Levees, by Rawle O. King.

Both The National Map and NILS represent federal efforts to foster interagency sharing of data into a single product providing national coverage of topography and federal land holdings respectively. The utility of both efforts is limited by the quality, accuracy, and completeness of the underlying geospatial data. The National Map, as currently envisioned, will provide topographic information at the 1:24,000 scale, meaning that roughly one inch on the map equals 2,000 feet. That scale will likely limit The National Map's usefulness for depicting, for example, floodplain boundaries to meet the requirements for FEMA floodplain maps.[28] Also, at some point in the future NILS presumably could provide one-stop shopping for an accurate assessment of the amount of federal land currently administered by each land management agency in the Department of the Interior and for the USFS. Currently, however, the best method for obtaining an accurate tally of federal lands is to contact each land management agency directly for their most up-to-date data.[29] (Even then, data from individual agencies do not always sum to totals reported by the General Services Administration.)[30]

Both NILS and The National Map are test cases for whether and how the agencies implement the OMB-issued supplemental guidance (discussed above) for managing geospatial assets as a portfolio of investments. For example, under the new guidance presumably both projects would receive some indication of where they stand in comparison to other geospatial dataset priorities. Further, it would be expected that under the new OMB guidance those priorities would be reflected in annual budget requests from the relevant agencies, which OMB would review and approve prior to their submission to Congress.

LEGISLATION IN THE 112TH CONGRESS

H.R. 1620/S. 1153, the Federal Land Asset Inventory Reform Act

One bill introduced in the 112th Congress on March 15, 2011, attempts to address some of the challenges to data coordination discussed above. The Federal Land Asset Inventory Reform Act of 2011 (H.R. 1620) would establish a national cadastre.[31] The legislation has been referred to the Subcommittee on Energy and Mineral Resources, House Natural Resources Committee, but has not been acted upon. A companion bill, S. 1153, was introduced on June 7, 2011, in the Senate and referred to the Senate Energy and Natural Resources Committee. It also has not been acted upon.

H.R. 1620 would require the Secretary of the Interior to develop a multipurpose cadastre of federal "real property." The legislation defines cadastre as an inventory, and defines federal "real property" as land, buildings, crops, forests, or other resources still attached to or within the land or improvements or fixtures permanently attached to the land or structures on it. The bill would require the Secretary to coordinate with the FGDC pursuant to OMB Circular A-16, to integrate the activities under the legislation with similar cadastral activities of state and local governments, and to participate in establishing standards and protocols that are necessary to ensure interoperability of the geospatial information of the cadastre for all users.[32]

Supporters of the legislation claim that a national cadastre would improve federal land management, resource conservation, environmental protection, and the use of federal real property. As noted above, the BLM currently has responsibility for maintaining federal land parcel information and ownership status, and it is not clear if H.R. 1620 would expand the current geospatial activities at BLM, shift the custodial responsibilities to another agency, or result in a different approach or program. Beginning with versions of these bills first introduced in the 110th Congress, supporters indicated that existing inventories of federal real property are old, outdated, and inaccurate, and updates to these inventories could improve management of the federal lands.[33] Observers also noted that the federal government lacks one central inventory that coordinates all the inventories into one usable database.[34]

H.R. 4233, The Map It Once, Use It Many Times Act

On March 21, 2012, Representative Lamborn introduced H.R. 4233, the Map It Once, Use It Many Times Act. In Title I, the bill would establish a National Geospatial Technology

Administration in the Department of the Interior (DOI) that would create a National Geospatial Database of all U.S.-owned or -managed lands, Indian trust lands, and nonfederal lands. Funding for the database would be at the discretion of the Administrator, who would develop and implement a funding strategy using appropriated funds, user fees, a revolving fund, partnerships with other government agencies, public-private partnerships, or some combination of some or all of the funding options.

The Administrator of the new organization would assume all geospatial functions currently vested by law in the DOI, the National Forest System within the U.S. Department of Agriculture, and NOAA, including all functions of the National Geodetic Survey.[35] All federal agencies would make their geospatial data available for inclusion in the National Geospatial Database, and the database would be available to the public.[36] Further, § 108(a)(3) of the bill would require that the Director of the Bureau of the Census make available, for inclusion in the database, all building addresses and geographical coordinates collected by the Census Bureau.[37] Currently the Census Bureau is forbidden to publish any private information—such as names, addresses, and telephone numbers.[38] (Privacy issues are discussed below under the section entitled "Some Privacy Issues: Census, Farm Bill, Internet.")

H.R. 4233 would also create a National Geospatial Policy Commission. The commission would produce a National Geospatial Data Plan, and the National Geospatial Technology Administration would be required to carry out the recommendations in the plan. The commission would be charged with coordinating federal agencies, state and local governments, and private entities to "eliminate redundancy in the performance of geospatial activities."[39] The bill also calls on the commission to direct geospatial activities to private-sector firms when possible.

Two of the five titles in H.R. 4233 deal with moving geospatial activities to private-sector firms and encouraging federal use of private geospatial firms generally.[40] Title III lays out requirements for contractor performance of geospatial activities, and Title IV calls for developing a strategy to encourage and enhance the use of private geospatial firms by federal agencies and by other organizations that receive federal funds. Title V of the bill contains provisions for encouraging geospatial research and development.

LEGISLATION IN THE 111TH CONGRESS

The Ocean and Coastal Mapping Integration Act

The Ocean and Coastal Mapping Integration Act, introduced as S. 174 and H.R. 365 in the 111th Congress, was enacted into law as Subtitle B of Title XII of the Omnibus Public Land Management Act of 2009 (P.L. 111-11). The act established a federal program to develop a coordinated and comprehensive mapping plan for the coastal waters including the exclusive economic zone and continental shelf, and the Great Lakes. Programs established by the act are intended to address issues of data sharing and cost-effectiveness by fostering cooperative mapping efforts, developing appropriate data standards, and facilitating the interoperability of data systems. Further, the program established under the act would develop these standards to be consistent with the requirements of the FGDC, so that the data collected in support of mapping are useful not only to the federal government, but also to coastal states and other entities. The theme of coordinating activities is underscored in several places in the act, specifically with other federal efforts such as the Digital Coast,[41] Geospatial One-Stop,[42] and the FGDC, as well as international mapping activities, coastal state activities, user groups, and nongovernmental entities.

Subtitle B of Title XII of P.L. 111-11 called for a plan for an integrated ocean and coastal mapping initiative within NOAA. The agency submitted the plan to Congress on May 24, 2010.[43] A biennial progress report on implementing the subtitle, also called for in P.L. 111-11, from the Interagency Committee on Ocean and Coastal Mapping is undergoing interagency review and is not available.[44]

The challenge to collect and manage the geospatial data needed to meet the requirements of the act is daunting, given the array of federal agencies, affected states, local communities, businesses, and other stakeholders who have an interest in coastal mapping. Moreover, the stakeholders require wide and disparate types of data—such as living and nonliving coastal and marine resources, coastal ecosystems, sensitive habitats, submerged cultural resources, undersea cables, aquaculture projects, offshore energy projects, and others. Congress could view the development of the ocean and coastal mapping plan and its implementation as a test case: how to manage a large data collection effort—cost-effectively and cooperatively—that reaches across all levels of government and includes interest groups, businesses, NGOs, and even international partners.

NONFEDERAL STAKEHOLDERS

Nonfederal organizations and institutions have increasingly participated with federal agencies in communicating their concerns regarding geospatial information management, data sharing, and coordination. The National Geospatial Advisory Committee directly advises the FGDC. The National States Geographic Information Council and private sector geospatial organizations, such as the Management Association for Private Photogrammetric Surveyors, provide views of state geospatial organizations and the private sector respectively.

National Geospatial Advisory Committee

The National Geospatial Advisory Committee (NGAC) was formed in early 2008 to provide advice and recommendations to the FGDC on management of federal geospatial programs, development of the NSDI, and implementation of the OMB Circular A-16. The committee members represent the private sector, nonprofits, academia, and governmental agencies.[45]

As part of its charter, NGAC provides a forum to convey views representative of nonfederal stakeholders in the geospatial community.

In its January 2009 report, *The Changing Geospatial Landscape*, NGAC noted that as geospatial data production has shifted from the federal government to the private sector and state and local governments, new partnerships for data sharing and coordination are needed. Specifically:

> the hodgepodge of existing data sharing agreements are stifling productivity and are a serious impediment to use even in times of emergency.... When the federal government was the primary data provider, regulations required data to be placed in the public domain. This policy jump-started a new marketplace and led to the adoptions of GIS capabilities across public and commercial sectors. However, these arrangements are very different when data assets are controlled by private companies or local governments.[46]

NGAC observed further that the federal government's need for land parcel (cadastral) data, which is also emphasized by the National Research Council, is missing an arrangement for acquiring the detailed property-related data necessary to make decisions during times of emergency. The report suggests that detailed land parcel data—its use, value, and ownership—is needed by FEMA, the USFS, and the U.S. Department of Housing and Urban Development for emergency preparedness, response to hurricanes and wildfires, or to monitor the current foreclosure problems.[47]

NGAC Recommendations to the Obama Administration

In October 2008, NGAC sent recommendations to the 2008-2009 Presidential Transition Team for improving the federal role in coordinating geospatial activities, for making changes to the *U.S. Code* pertaining to non-sensitive address data, and for enhancing geospatial workforce education.[48]

Most recommendations pertained to how the federal government could better coordinate geospatial partnerships with state, local, and tribal governments, the private sector, and the academic community. Recommendations included:

- establish a geospatial leadership and coordination function immediately within the Executive Office of the President; the geospatial coordination function should be included in the reauthorization of the E-Government Act;
- require OMB and FGDC to strengthen their enforcement of OMB Circular A-16 and EO 12906;
- establish/designate Geographic Information Officers with each department or agency with responsibilities stipulated within OMB Circular A-16;
- establish and oversee an Urgent Path[49] forward for implementation of geospatial programs necessary to support current national priorities and essential government services underpinning the NSDI; and
- continue NGAC.

Arguably, the supplemental guidance to OMB Circular A-16, issued on November 10, 2010, and the appointment of Vivek Kundra as the Federal Chief Information Officer at OMB address aspects of the first two bullets. OMB's memorandum issued on March 3, 2006, arguably should have addressed the third bullet; however, NGAC apparently felt that OMB's guidance had not been implemented.

Of the final two bullets, NGAC has continued to exist, and it is not clear what progress has been made in implementing NGAC's Urgent Path forward (one aspect of the Urgent Path, Imagery for the Nation, is discussed below).

National States Geographic Information Council (NSGIC)

At the national level, the FGDC exists to promote the coordinated development, use, sharing, and dissemination of geospatial data. At the state level, NSGIC[50] exists to promote the coordination of statewide geospatial activities in all states, and to advocate for the states in national geospatial policy initiatives to help enable the NSDI.[51] NSGIC ties its activities to the NSDI by promoting the development of Statewide Spatial Data Infrastructures (SSDI), under a partnership called the 50-States Initiative, which is intended to lead to the creation of an SSDI for each state. In this vision, each state's SSDI would enable coordination between geospatial data producers and consumers at all levels within the state, and allow the state to share geospatial data with the national geospatial structure envisioned as the NSDI. The emphasis on organization and coordination of geospatial data and activities is seen as critical to reducing costs to states and the federal government by eliminating data redundancy—collecting the data once, using it many times—and by setting standards that allow different users to share geospatial data regardless of who collects it.

NSGIC identified 10 criteria that define a "model" state program necessary to develop effectively coordinated statewide GIS activities, and thus reduce inefficiency and waste. These include:

1. strategic and business plans
2. a full-time, paid, GIS coordinator and staff
3. clearly defined authority and responsibility for coordination
4. a relationship with the state chief information officer
5. a political or executive champion for coordinating GIS
6. a tie to the national spatial data infrastructure and clearinghouse programs
7. the ability to work with local governments, academia, and the private sector
8. sustainable funding, especially for producing geospatial data
9. the authority for the GIS coordinator to enter into contracts, and
10. the federal government working through the statewide coordinating body

Not all states have fully embraced the need for statewide coordination of GIS activities, and states differ in their structure and organization of geospatial data among and between state, county, and local entities. For example, some states such as Arkansas share geospatial data across agencies in a very open manner; other states such as New York require more formal agreements or have restrictions to sharing data that include critical infrastructure. Nonetheless, some level of data sharing does occur, even in the more restrictive states.[52]

Imagery for the Nation—A Priority for NSGIC

A priority for NSGIC is a program under development, called Imagery for the Nation (IFTN), that would collect and disseminate aerial and satellite imagery in the form of digital orthoimagery.[53] In its description of the program, NISGIC notes that digital orthoimagery is the foundation for most public and private GIS endeavors. Further, NSGIC states that as many as 1,300 different government entities across the nation are developing digital orthoimagery products, " ...leading to higher costs, varying quality, duplication of effort, and a patchwork of products."[54] IFTN represents an effort to establish one coherent set of geospatial data—arguably one of the most important layers in a GIS, orthoimagery—that is organized for the benefit of many stakeholders at the federal, tribal, regional, state, and local levels.

As proposed, IFTN would involve two federal programs: (1) the existing National Agricultural Imagery Program (NAIP) administered by the U.S. Department of Agriculture, and (2) a companion program administered by the USGS. The NAIP imagery would be enhanced to provide annually updated one-meter resolution orthoimagery over all states except Hawaii and Alaska.[55] The USGS program would also collect one-foot resolution imagery every three years for 50% of the U.S. land mass (except Alaska, which would get one-foot resolution imagery only over densely populated areas). The program would include an option for states to "buy up," or enhance, any or all of the remaining 50%. The program would also provide 50% matching funds for partnerships to acquire six-inch resolution imagery over urban areas with at least 1,000 people per square mile as identified by the U.S. Census Bureau. NSGIC states that statewide GIS coordination councils would specify their requirements through business plans, and that all the data would remain in the public domain, which would address many of the data sharing issues discussed above. In addition, the program calls for appropriate national standards for all data, which is a goal of the FGDC, a partner to NSGIC in the development of IFTN. NSGIC estimates that the program would cost $1.38 billion during the first 10 years, and argues that this would save $120 million over the 10-year period by reducing the number of contracts, contracting for larger areas, reducing overhead, and reducing other costs associated with current efforts.[56]

Advancing the National Spatial Data Infrastructure: The NSGIC Perspective

NSGIC considers the 50-States Initiative as one of the crucial components needed to build the NSDI and to bring consistency of geospatial information and parity to each of the states.[57] NSGIC also considers that IFTN is the first of several initiatives creating "core data layers," or baseline data programs, required to meet federal, state, and local needs.[58] NSGIC suggests that the NGAC be an interim step in the governance structure for NSDI, and indicates that the national effort to govern and coordinate the geospatial enterprise should not stifle the states from customizing aspects of the NSDI to suit their own needs:

> the federal government must not dictate the actions of state and local governments, nor should state governments dictate those of local government. However, each level of government can exert a strong influence on subordinate levels by making funding contingent on compliance with the policies and standards it establishes.[59]

NSGIC further argues that federal funding for the NSDI could be modeled on the federal highway program, similarly contingent upon compliance with collaboratively established criteria and requirements.[60]

Some Privacy Issues: Census, Farm Bill, Internet

The Census and the Farm Bill

The Census Bureau is forbidden to publish any private information—such as names, addresses, and telephone numbers.[61] This type of geospatial information is available in the public domain for some localities[62] in the United States; however, it is not provided by the Census Bureau. In its recommendations, NGAC calls for revising "restrictive statutory language as it pertains to non-sensitive address data in Title 13 U.S. Code and to 'geospatial' data in Section 1619 of the 2008 farm bill."[63] Title 13 contains provisions for not disclosing or publishing private information that identifies an individual or business (Sections 9 and 214). A proposal to amend portions of Title 13 and make geospatial data collected by the Census Bureau more accessible will likely raise issues about the privacy of personal data collected by the federal government; the value of such data for emergency management; disaster preparation; other local, regional, and national needs; and the various tradeoffs between privacy concerns and the accessibility to geospatial data. As discussed above, legislation introduced in the 112th Congress—H.R. 4233—contains provisions to make geospatial Census data publicly available, notwithstanding Title 13 of the *U.S. Code*.

Section 1619 of the 2008 farm bill (P.L. 110-246) prohibits disclosure of geospatial information about agricultural land or operations, when the information is provided by an agricultural producer or owner of agricultural land and maintained by the Secretary of Agriculture. Certain exceptions, contained in Section 1619 of the 2008 farm bill, apply to the prohibition. NGAC has taken the position that the statutory language could be revised to enhance the value of the geospatial data while not compromising privacy.[64]

Internet Privacy

Concerns about access to geospatial information arose following the release of a draft Federal Trade Commission (FTC) report on Internet privacy.[65] The report focused on "behavioral advertising," whereby advertisements are tailored to web users based on data collected about them across the web. The report proposed that Internet companies create a "do-not-track" mechanism that would allow Internet users to opt out of online data collection.[66] Although the thrust of the report was not focused on geospatial information, some private-sector interests raised concerns about how access to geospatial information might be affected under a "do-not-track" mechanism. The Management Association for Private Photogrammetric Surveyors (MAPPS), an association of private geospatial firms,[67] expressed its concern with the term "precise geolocation data," which MAPPS noted was undefined in the FTC report. In a letter to the FTC, MAPPS wrote: "The use of the term 'geolocation' or other geospatial relevant terminology, as it is used in this or possible future FTC regulation, could thwart legitimate and desirable business activities; deny consumers the products, technologies and services they are demanding in the marketplace; and impose a significant new liability on our members."[68] In its letter, MAPPS requested that the FTC either remove any reference to the term "precise geolocation data," define the term, or exempt firms from having to obtain consent from consumers before collecting the geospatial data. MAPPS added that "it would be impractical, if not impossible, for our member firms to obtain prior approval or consent from individual citizens prior to acquiring or applying data such as satellite imagery, aerial photography, or parcel, address, or transportation data."[69]

On March 26, 2012, FTC issued the final report.[70] The report included a footnote that provided some clarification over the issue of collecting geolocation data for the purposes of mapping: "With respect to use of geolocation data for mapping, surveying, or similar purposes, if the data cannot reasonably be linked to a specific consumer, computer, or device, a company collecting or using the data would not need to provide a consumer choice mechanism."71 Although the footnote seems to address the issue raised by MAPPS (as discussed above), its interpretation likely hinges on how the term "reasonably" is applied.

A NATIONAL GIS?

In early 2009, several proposals were released calling for efforts to create a national GIS,[72] or for renewed investment in the national spatial data infrastructure, or even to create a "NSDI 2.0."[73] The language in the proposals attempted to make the case for considering such investments part of the national investment in critical infrastructure, both by directly supporting these national GIS and geospatial efforts, but also via secondary effects. For example, one proposal indicated that organizations rebuilding roads, bridges, and schools need updated online information networks "to rebuild in a smart, efficient, environmentally conscious and sustainable way."[74] Another proposal touted a national GIS as a tool to speed economic recovery, which could also "leave the country with a public utility, a modern geospatial information system, that itself can become a foundation for new generations of industries and technologies in the future."[75]

Their call for efforts to build a "national" GIS, or a new version of the NSDI, or for an investment in a national spatial data infrastructure, raises questions about the current efforts to build the NSDI. Efforts to construct the NSDI began in 1994 with Executive Order 12906, or even earlier when OMB revised Circular A-16 in 1990 to establish the Federal Geographic Data Committee. The recent proposals imply that efforts which began 20 years ago and continue today are not sufficiently national in scope, planning, coordination, sharing, or implementation, despite the existence of the FGDC, NSGIC, or other entities such as the Coalition of Geospatial Organizations or MAPPS that are forums for organizations concerned with national geospatial issues.

CONCLUSION

Congress may consider how a national GIS or geospatial infrastructure would be conceived, perhaps drawing on proposals for these national efforts as described above, and how they would be similar to or differ from current efforts. Congress may also examine its oversight role in the implementation of OMB Circular A-16, particularly in how federal agencies are coordinating their programs that have geospatial components. In 2004, GAO acknowledged that the federal government, through the FGDC and Geospatial One-Stop project, had taken actions to coordinate the government's geospatial investments, but that those efforts had not been fully successful in eliminating redundancies among agencies. As a result, federal agencies were acquiring and maintaining potentially duplicative data sets and systems.[76] Since then, it is not clear whether federal agencies are successfully coordinating among themselves and measurably eliminating unnecessary duplication of effort.

Were Congress to take a more active oversight role overseeing the federal geospatial enterprise, it could evaluate whether specific recommendations from nonfederal stakeholders have been addressed. For example, the National Geospatial Advisory Committee recommended that OMB and FGDC strengthen their enforcement of Circular A-16 and

Executive Order 12906. However, enforcement alone may not be sufficient to meet the current challenges of management, coordination, and data sharing. The issuance of supplemental guidance to Circular A-16 by OMB in November 2010 may instigate new activity among and between agencies, which could spill over into better coordination with the state and local governments and the private sector.

It will likely take some time, and several budget cycles, to track whether agencies are adhering to the "portfolio-centric model" of geospatial data management outlined in the supplemental guidance. It may also take time to evaluate whether the "portfolio-centric model" is the best available model for managing the federal geospatial assets.

APPENDIX. HISTORY AND BACKGROUND OF CIRCULAR A-16[77]

This Circular was originally issued in 1953, revised in 1967, and revised again in 1990. The Bureau of the Budget (now the OMB) issued Circular No. A-16 on January 16, 1953. Appended to this Circular were Exhibits, occasionally revised, that dealt with procedures for programming and coordinating of federal Topographic Mapping Activities, National Atlas, Geodetic Control Surveys and International Boundaries.

The purpose of the 1953 Circular was "to insure (sic) that surveying and mapping activities may be directed toward meeting the needs of federal and state agencies and the general public, and will be performed expeditiously, without duplication of effort." The original Circular references Executive Order No. 9094, dated March 10, 1942. This Executive Order directs the Director of the Bureau of the Budget to coordinate and promote the improvement of surveying and mapping activities of the Government. Furthermore, it passes on functions carried out by the Federal Board of Surveys and Maps, established by Executive Order No. 3206, dated December 30, 1919. Thus, the OMB is directed to make recommendations to agencies and to the President regarding the coordination of all governmental map making and surveying. Executive Order No. 3206 superseded an Executive Order, dated August 10, 1906, that granted advisory power to the United States Geographic Board to review mapping projects to avoid duplication and to facilitate standardized mapping.

A revised Circular A-16 was issued on May 6, 1967. The most significant change in this revision is the addition of a new section on Responsibility for Coordination. This section outlines the responsibilities of three federal departments (Department of the Interior (DOI), Department of Commerce (DOC) and Department of State (DOS)). Both the original and the 1967 revision of the Circular focus on providing a guide for the development of annual programs of the individual agencies and, through the Exhibits, established extensive reporting requirements.

A second revised Circular A-16 was issued on October 19, 1990. This revision expanded the Circular to include not only surveying and mapping, but also the related spatial data activities. Specifically, it included geographically referenced computer-readable (digital) data. In addition, the Exhibits are no longer referenced and a short reporting requirements section is added.

The 2002 updated Circular calls for continued improvements in spatial data coordination and the use of geographical data. Objectives for this revision are to reflect the changes that have taken place in geographic information management and technology, and to clearly define agency and FGDC responsibilities. The proposed revision displays an integrated infrastructure system approach to support multiple government services and electronic government.

End Notes

[1] The National Geospatial Advisory Committee, The Changing Geospatial Landscape, January 2009, p. 9, http://www.fgdc.gov/ngac/NGAC%20Report%20-%20The%20Changing%20 Geospatial%20Landscape.pdf. Hereafter referred to as NGAC, The Changing Geospatial Landscape, January 2009.

[2] Prepared statement of Rep. Adam Putnam, Chair, U.S. Congress, House Committee on Government Reform, Subcommittee on Technology, Information Policy, Intergovernmental Relations and the Census, Geospatial Information: A Progress Report on Improving our Nation's Map-related Data Infrastructure, 108th Cong., 1st sess., June 10, 2003, H. Hrg. 108-99 (Washington: GPO, 2004).

[3] OMB Circular A-16 was originally issued in 1953; however, its lineage stretches back to the early part of the 20th century, and its focus on coordination and avoiding duplication began in an Executive Order dated August 10, 1906. A 1990 revision expanded the Circular beyond surveying and mapping, its original focus, to include digital data, such as the geospatial data discussed in this report. Appendix of this report reproduces Appendix C of Circular A-16, which describes its history and background.

[4] Executive Order 12906, "Coordinating Geographic Data Acquisition and Access: The National Spatial Data Infrastructure," 59 Federal Register, April 13, 1994.

[5] The NSDI in its original definition under EO12906 means "a distributed network of geospatial data producers, managers, and users linked electronically."

[6] The FGDC is an interagency committee charged with coordinating development, use, sharing, and dissemination of geospatial data on a national basis. Members include representatives from the Executive Office of the President, and Cabinet-level and independent federal agency representatives. See http://www.fgdc.gov/.

[7] Vivek Kundra, Federal Chief Information Officer, Geospatial Line of Business OMB Circular A-16 Supplemental Guidance, Office of Management and Budget, November 10, 2010, http://www.whitehouse.gov/sites/ default/files/omb/ memoranda/2011/m11-03.pdf.

[8] Cited in U.S. General Accounting Office, Geographic Information Systems: Challenges to Effective Data Sharing, GAO-03-874T, June 10, 2003, p. 5. Hereafter referred to as GAO (2003). The 2006 Annual Report from the Federal Geographic Data Committee claims that 80%-90% of government information has a spatial component. (GAO became the Government Accountability Office effective July 7, 2004.)

[9] Recognition of the inherent propensity for duplication of effort arguably was first addressed by an Executive Order issued on August 10, 1906. That order allowed the United States Geographic Board (created by the Executive Order) to review mapping projects to avoid duplication and to facilitate standardized mapping. OMB Circular A-16, Appendix C. (See Appendix of this report for the text of Circular A-16, Appendix C.)

[10] Per Executive Order 12906.

[11] Milo Robinson, A History of Spatial Data Coordination, Federal Geographic Data Committee, A white paper available via the National Geospatial Advisory Committee, May 2008, http:// www.fgdc.gov/ngac/a-history-of-spatialdata-coordination.pdf.

[12] Clay Johnson III, OMB Deputy Director for Management, Designation of a Senior Agency Official for Geospatial Information, OMB Memorandum for Heads of Selected Executive Departments and Agencies, March 3, 2006, http://www.whitehouse.gov/sites/default/files/ omb/assets/omb/memoranda/fy2006/m06-07.pdf.

[13] Vivek Kundra, Federal Chief Information Officer, Geospatial Line of Business OMB Circular A-16 Supplemental Guidance, Office of Management and Budget, November 10, 2010, http://www.whitehouse.gov/sites/ default/files/omb/ memoranda/2011/m11-03.pdf.

[14] OMB Circular A-16, Supplemental Guidance, p. 5.

[15] OMB Circular A-16, Supplemental Guidance, pp. 3-4.

[16] OMB Circular A-16, Supplemental Guidance, p.4.

[17] OMB Circular A-16, Supplemental Guidance, p. 12.

[18] Ibid., p. 28.

[19] For more information about land parcel data, see CRS Report R40717, Issues Regarding a National Land Parcel Database, by Peter Folger.

[20] National Research Council, National Land Parcel Data: A Vision for the Future, Washington, DC, 2007, p. 2. Hereafter referred to as NRC, National Land Parcel Data.

[21] NRC, National Land Parcel Data.

[22] NRC, National Land Parcel Data, p. 7.

[23] The National Map would be the next generation of topographic maps—online and interactive—that would supplant the paper versions that the USGS has produced for decades. The National Map is envisioned as a consistent framework for geographic knowledge nationwide. For more information, see CRS Report R41825, Geospatial Information and Geographic Information Systems (GIS): An Overview for Congress.

[24] GAO (2003), p. 6.

[25] Circular A-16, at http://www.whitehouse.gov/omb/circulars/a016/a016_rev.html.

[26] The Department of Defense also administers a significant amount of land.

[27] See http://www.geocommunicator.gov/GeoComm/index.shtm.

[28] One report notes that the USGS National Elevation Dataset (NED) is commonly used in flood map production, even though the uncertainties inherent in the elevation data are about 10 times greater than those defined by FEMA as acceptable for flood plain mapping. See National Research Council, Mapping the Zone: Improving Flood Map Accuracy, Washington, DC, 2009, p. 3.

[29] E-mail from John P. Donnelly, National Atlas of the United States, USGS, Reston, VA, February 4, 2009.

[30] General Services Administration, FY 2009 Federal Real Property Statistics, Published by the GSA Office of Governmentwide Policy, 2009, http://www.gsa.gov/graphics/ogp/ FY2009_ FRPR_Statistics.pdf.

[31] A cadastre is a map or database of ownership and boundaries of land parcels.

[32] Similar legislation (H.R. 1520) was introduced in the 111th Congress but was not enacted. In the 110th Congress, H.R. 5532 and S. 3043 were introduced but neither version of the bill saw action.

[33] "Legislators return with FLAIR," GEO World, May 2008, p. 15.

[34] Ibid.

[35] The National Geodetic Survey mission is "to define, maintain and provide access to the National Spatial Reference System to meet our nation's economic, social, and environmental needs." See http://www.ngs.n oaa.gov/INFO/ WhatWeDo.shtml.

[36] Except for information that could cause damage to U.S. national security, per § 103 (C)(2) of the bill.

[37] That section of the bill also states that these data would be provided for the National Geospatial Database "to the extent consistent with individual privacy protections."

[38] 13 U.S.C. § 9 and § 214. See also U.S. Census Bureau, at http://www.census.gov/privacy/ data_protection/ federal_law.html.

[39] Title II, § 201(2) of H.R. 4233.

[40] Titles III and IV of H.R. 4233.

[41] The Digital Coast is a NOAA-led effort envisioned as a an information delivery system for coastal data, as well as the training, tools, and examples needed to turn data into useful information. See http://www.csc.noaa.gov/ digitalcoast/ index.html.

[42] Geospatial One-Stop is the means of accessing metadata resources published through the National Spatial Data Clearinghouse and managed in NSDI.

[43] E-mail from Michael Jarvis, Congressional Affairs Specialist, NOAA, May 17, 2011. The report is available at http://www.iocm.noaa.gov/reports/2010 _NOAA_OCMIA_ Report_ to_Congress.pdf.

[44] E-mail from Michael Jarvis, May 17, 2011.

[45] National Geospatial Advisory Committee membership, http://www.fgdc.gov/ ngac/ member ship. The committee is sponsored by the Department of the Interior under the Federal Advisory Committee Act.

[46] NGAC, The Changing Geospatial Landscape, January 2009, p. 12.

[47] Ibid.

[48] See http://www.fgdc.gov/ngac/ngac-transition-recommendations-10-16-08.pdf.

[49] The NGAC recommendations further specify that an "Urgent Path" forward should include (1) Imagery for the Nation; (2) National Land Imaging Program; and (3) National Land Parcel data.

[50] Members of NSGIC include senior state GIS managers and coordinators, although other members include representatives from federal agencies, local government, the private sector, academia, and other professional organizations. See http://www.nsgic.org/ about /index.cfm.

[51] National States Geographic Information Council Strategic Plan 2009-2011, at http://www. nsgic.org/resources/ strategicplan.pdf.

[52] E-mail from Learon Dalby, NSGIC President 2008-2009, March 11, 2009.

[53] Orthoimagery is an aerial or satellite image or photograph from which distortions resulting from camera tilt and ground relief have been removed. An orthophoto or orthoimage has a uniform scale and can be used as a map.

[54] See NISGIC, Imagery for the Nation, at http://www.nsgic.org/hottopics /iftn_brochure_ 0308. pdf.

[55] Imagery would be updated once every three years in Hawaii. The USGS program would produce one-meter imagery for Alaska once every five years.

[56] CRS did not review the basis for NSGIC's cost analysis, nor examine the cost benefit analysis completed for the IFTN in July 2007.

[57] NSGIC, A Strategic Framework for the National Spatial Data Infrastructure, http://www. nsgic.org/resources/ strategic_framework_NSDI_NSGIC.pdf.

[58] Ibid; for example, NSGIC suggests that Imagery for the Nation should probably be followed by Elevation for the Nation, Transportation for the Nation, Cadastral for the Nation, and so on.

[59] NSGIC, The States' Perspective on Advancing the National Spatial Data Infrastructure, October 10, 2008.

[60] NSGIC, A Strategic Framework for the National Spatial Data Infrastructure, p. 19.

[61] 13 U.S.C. § 9 and §214. See also U.S. Census Bureau, at http://www.census.gov/ privacy/ data_protection/ federal_law.html.

[62] For example, the website for the City of Greeley, CO, property information map, identifies names and addresses, the underlying street map and orthoimagery, together with other information such as school districts and even the nearest fire hydrant. See http://gis. greeleygov.com/origin/propinfo.html.

[63] In Title 13 Congress delegates responsibility for conducting the Census to the Secretary of Commerce.

[64] Telephone conversation with Anne Miglarese, Chair, National Geospatial Advisory Committee, May 26, 2009.

[65] Federal Trade Commission, Protecting Consumer Privacy in an Era of Rapid Change: A Proposed Framework for Businesses and Policy Makers, Preliminary FTC Staff Report, Washington, DC, December 2010, http://www. ftc.gov/os/2010/12/101201 privacyreport. pdf.

[66] See Wendy Davis, "FTC Criticizes Self-Regulatory Efforts, Proposes Blanket Do-Not-Track," MediaPostNews, Online Media Daily, December 1, 2010, http://www.mediapost.com/ publications/?fa=Articles.show Article&art_aid= 140435.

[67] The Executive Director for MAPPS, John M. Palatiello, has served on NGAC.

[68] Letter to the Honorable Jonathan D. "Jon" Liebowitz, Chairman, Federal Trade Commission, from John M. Palatiello, Executive Director, The Management Association for Private Photogrammetric Surveyors, January 4, 2011, http://www.mapps.org/issues/MAPPS_ Letter_to_FTC_1-4-11.pdf. The FGDC Secretariat also commented on the preliminary report, echoing MAPPS' concerns about the term "precise geolocation data." See http://www.ftc.gov/os/comments/privacyreportframework/00354-57961.pdf.

[69] MAPPS letter to the FTC.

[70] Federal Trade Commission, Protecting Consumer Privacy in an Era of Rapid Change: Recommendations for Businesses and Policymakers, March 2012. Available at http://ftc. gov/opa/2012/03/privacyframework.shtm.

[71] FTC, Protecting Consumer Privacy in an Era of Rapid Change, footnote 187, p. 39.

[72] These proposals are broader than what is currently envisioned as The National Map, under the USGS.

[73] See, for example, the following: A Proposal for National Economic Recovery: An Investment in Geospatial Information Infrastructure Building a National GIS, at http://www. gis.com/gisnation/pdfs/ national_ economic_recovery.pdf; A Proposal for Reinvigorating the National Economy Through Investment in the US National Spatial Data Infrastructure, at http://www.cast.uark.edu/nsdi/nsdiplan.pdf; and A Concept for American Recovery and Reinvestment, NSDI 2.0: Powering our National Economy, Renewing our Infrastructure, and Protecting our Environment, at http://www.nsdi2.net/NSDI2 ProposalFor American RecoveryAndReinvestment_V1_4.pdf.

[74] A Concept for American Recovery and Reinvestment, NSDI 2.0: Powering our National Economy, Renewing our Infrastructure, and Protecting our Environment, p. 2.

[75] A Proposal for National Economic Recovery: An Investment in Geospatial Information Infrastructure Building a National GIS.

[76] GAO (2004), p. 19.

[77] Circular A-16 Revised, White House Office of Management and Budget, Appendix C. Reproduced here in its entirety.

In: Federal Geospatial Information
Editors: Andrew Halpern and Mark Troupp ISBN 978-1-62257-572-5

Chapter 3

MODERNIZATION ROADMAP FOR THE GEOSPATIAL PLATFORM*

U.S. Department of the Interior
Federal Geographic Data Committee

1. FOREWORD

We are pleased to share version 4 of the Modernization Roadmap for the Geospatial Platform (Roadmap). This final version of the Roadmap follows a stakeholder engagement process intended to share the vision for the Platform and gather feedback from the geospatial community regarding concepts discussed in the Roadmap. Feedback on the Roadmap to date has been positive, while at the same time identifying some challenging issues and questions that will be addressed as implementation occurs.

Developing the Roadmap has been an evolving process that began in February 2010 with the release of the President's 2011 Budget, which provided initial direction to develop the Geospatial Platform. An earlier draft Roadmap (version 3), was approved by the Office of Management and Budget (OMB) on November 10, 2010. That document, prepared by member agencies of the Federal Geographic Data Committee (FGDC), received broad Federal review and upon OMB approval set the stage for a more robust stakeholder engagement process.

The geospatial community at large has been engaged in providing feedback on the Roadmap over the past several months through a variety of mechanisms that include:

- Review and feedback from the National Geospatial Advisory Committee (NGAC), a Federal Advisory Committee that provides advice to the Federal government on geospatial matters;
- Presentations and discussions at geospatial conferences and related professional meetings; and

---

* This is an edited, reformatted and augmented version of the US Department of the Interior, Federal Geographic Data Committee publication, dated March 2011.

- Development of an IdeaScale Web-based forum for gathering input and feedback through the Geospatial Platform Web page_(http://www.geoplatform.gov).

These outreach mechanisms resulted in valuable feedback that can be broadly categorized into questions and comments related to the following areas:

- Policy and management;
- Governance;
- Lessons to be learned from other model programs, and;
- Clarification of concepts.

The Geospatial Platform team has listened to the insights, ideas, and concerns expressed by geospatial stakeholders and offer the following summary of the approach for addressing each these areas of feedback as the Geospatial Platform moves into implementation throughout 2011 and beyond. This final Roadmap document does not fully address all of the issues identified through the stakeholder engagement process. Some of the issues raised through the outreach process will be addressed through the implementation process for the Platform.

Policy and Management

As the Geospatial Platform begins to take shape there are uncertainties about issues that may include authorities to operate the Platform, the selection of a Managing Partner and its roles and responsibilities, and the funding model. These and other critical issues identified during this initial period are difficult to address adequately in a planning document like the Roadmap. Throughout 2011, the Geospatial Platform leaders will develop specific position papers and strategies to address concept of operations, a business plan, and other relevant guidance documents to steer Platform implementation.

Governance

Comments on the Roadmap identified questions about the approach for governing the Geospatial Platform. We envision a governance process that includes representation from both Federal and non-Federal stakeholders. The Geospatial Platform leadership and the NGAC Intergovernmental Subcommittee are working to develop an inclusive governance model while the identification, roles, and responsibilities of the Managing Partner are being developed by executive leadership of the FGDC.

Lessons to Be Learned from Other Model Programs

A number of existing programs have been suggested as successful models that can offer insights to assist the implementation of the Platform. These programs often share common elements that are important for sustainable participation of partners and stakeholders. Organizational models and programs identified through the stakeholder engagement efforts are currently being evaluated to document best practices and lessons learned, and will be considered for applicability for the Geospatial Platform.

Clarification

Version 4 of the Roadmap attempts to incorporate ideas and feedback that result in a more clear presentation and description of the ideas contained in the Roadmap. As implementation of the Geospatial Platform occurs, components (or —pillars||) that appear independent in the Roadmap will become more closely interrelated, providing a dynamic environment that will continue to grow and change as information is gathered, the offering is developed, and lessons are learned.

Next Steps

Version 4 of the Roadmap provides a framework document that describes the concepts and goals of the Geospatial Platform. This final version concludes the upfront planning process for the Platform and paves the way for implementation activities.

Over the past several months, the Federal geospatial community has made a number of advances that positively affect implementation of the Geospatial Platform. While some of these issues are not fully addressed in the Roadmap, they will be integrated into future implementation plans for the Platform. Three examples include:

- As the FGDC Geospatial cloud computing test-bed initiative progresses, the lines between common services and shared infrastructure become softer as Infrastructure as a Service (IaaS) becomes a reality.
- OMB's release in November 2010 of the OMB Circular A-16 Supplemental Guidance (A-16 Supplemental Guidance) provides a method for implementing a geospatial portfolio management process for Federal geospatial data investments.
- The Administration has enhanced Data.gov as a public capability for citizens, business, and governmental agencies to gain access to government data. Geospatial data comprise the majority of the data in Data.gov, and based on public feedback of the importance of place-based data and visualization, Data.gov and Geospatial One-Stop (GOS) are being integrated.

In addition, the time boxes found in version 3 of the Roadmap have been removed from version 4. While the time boxes provided an initial course of specific actions, the next versions of tasks and deliverables for the Platform will be documented in implementation plans.

In the future, the concepts and approach put forth in the Roadmap will be further clarified and refined by supplemental documents that may include, but are not limited to;

implementation plans, business plans, technical documents, project plans, and best practice examples.

In closing, the Platform team thanks all who shared ideas and comments, as well as those who put in countless hours to develop, communicate, and support the Geospatial Platform. The input from the geospatial community has been invaluable in developing the Platform approach; and we look forward to continued dialogue and ongoing involvement of our stakeholders as the Geospatial Platform moves into its next phase of development.

2. PREAMBLE

The Need for Geospatial Capabilities

The challenges this Nation faces are multifaceted and complicated; they require diverse information, innovative analysis, and collaboration to solve. These challenges also have one other common characteristic – they occur in places.

Consider some of the most devastating events in recent United States history: the terrorist attacks on September 11, 2001, Hurricane Katrina, the mortgage crisis, and the Deepwater Horizon oil spill in the Gulf of Mexico. Internationally, consider the earthquake in Haiti. Responses to such events require national and international coordination, information sharing and analysis, and deployment of operational capabilities, all of which rely on knowledge of place.

Some complex questions raised by such events are evident in last year's Gulf oil spill:

- How can we best see, understand, track, and monitor the existing, pending, and future impacts of this event?
- Where do we mobilize and apply resources to provide the greatest good, prevent the greatest harm, and protect the life, health, and welfare of citizens?
- How do we synthesize vast information assets of the Federal, State, regional, local, and Tribal governments, the private sector, academia, and citizens to provide the on-the-ground situational awareness required to make wise decisions?
- How do we determine where proactive steps can be taken to prevent or minimize future impact such as identifying Gulf Coastal communities where aid will be required?

In addition to responding to emergencies, many of the day-to-day activities of government and citizens rely on information linked to places and communities. Geospatial capabilities and analyses help government decision makers target agency investments, measure progress towards program goals and measure performance across regions or communities. In addition, displaying results through interactive maps helps communicate complex ideas clearly. For citizens, geospatial capabilities and analyses help answer questions and solve problems related to commuting to work, choosing where to live and raise families, protecting our natural resources, planning vacation destinations, and juggling busy schedules, because all require knowledge about place.

Using place as an information management framework allows decision makers to leverage government databases, visualization techniques, and geospatial analysis tools to rapidly recognize patterns in large amounts of data and make better-informed decisions. This "place-based" framework is consistent with Administration Priorities for place-based budgeting[1] and society‘s increasing demand for and reliance on geo-enabled products and services.

The Case for the Geospatial Platform

Geospatial assets are already an integral part of many government and private day-to-day operations. While the business needs of stakeholders vary, there are many instances where different operations require the use of similar assets. Because geospatial information often involves a significant investment of resources (i.e., human, financial, etc.), many governments already coordinate their efforts to produce cost-savings, improve quality of services, and increase efficiency, although more could be done.

Coordinating the development and delivery of the Nation‘s geospatial assets is a complicated task, but the geospatial community has come together to create an initiative that does just that. The Geospatial Platform is the culmination of these efforts. The partner agencies of the FGDC are developing the foundational plans for the Geospatial Platform to leverage the geospatial expertise, and existing geospatial data, services, applications, and infrastructure to enhance the Nation‘s management of—and access to—all things geospatial. It will promote sharing across the geospatial community and provide opportunities to view and conduct business in new and innovative ways.

Under the basic premise "build it once, use it many times," benefits already realized through coordination will be enhanced and expanded when implemented on a larger scale through the Geospatial Platform. The Platform will provide the means to enhance and expand these capabilities by offering an operational focal point for delivering access to trusted geospatial data, services, applications, and infrastructure.

The Platform will promote interoperable, effective and efficient problem solving tools spanning across organizations, thereby reducing the cost associated with developing independent geospatial capabilities while providing effective and timely information to decision makers.

3. Executive Summary

The Purpose of the Geospatial Platform

The Geospatial Platform (Platform), an initiative based in the Fiscal Year (FY) 2011 Presidential Budget, with the goal of —ultimately increasing access to geospatial data,"[2] is designed to become a component of the National Spatial Data Infrastructure (NSDI).

The Platform will offer access to trusted geospatial data, services, applications, and infrastructure assets. By delivering access to geospatial assets that are —built once and used many times," the Platform will increase information sharing across various levels of government and the private sector, allowing for the reuse and adaptation of geospatial assets. This multi-purposing and availability of assets can lead to cost-savings, wider use of geospatial capabilities, and higher quality assets.

The Geospatial Platform will fundamentally improve access to and management of geospatial assets through a focus on five key components, or Pillars:

- Common Data, Services and Applications.
- Shared Infrastructure.
- Segment Architecture.
- Governance.
- Portfolio Management.

The Geospatial Platform will build upon the successes of ongoing interagency geospatial initiatives by identifying and allying with programs that have resulted in (1) successful sharing of assets, (2) efficiencies in developing collaborative programs, and/or (3) development of a standard national dataset through intergovernmental coordination. The Platform will complement current Administration initiatives for openness and transparency of government information and data, such as Data.gov, that have led to new thoughts and innovative approaches for delivering Federal information technology (IT).

The Business Model

The Platform will utilize a Business Model that outlines the business institution, identifies potential customers, and begins to consider a financial approach for managing the Platform and its assets.

The Business Model places emphasis on managing a Partner Network of providers. Customers may include Federal agencies and their partners in State, local, regional and Tribal governments, non-profit organizations, academic institutions, industry, and citizens. The Business Model approach represents a fundamental shift from managing geospatial assets on a volunteer or ―other duties as assigned‖ basis to creating sustainable support for a Managing Partner.

Implementation and Performance Goals

In the coming months, the Managing Partner will engage in a number of foundational activities as part of the Geospatial Platform‘s phased implementation.

The path forward includes ongoing collaboration with stakeholders and leveraging select, existing geospatial efforts in all levels of government and the private sector.

To assess how well the Platform meets its goals and creates value for customers, the Roadmap outlines preliminary performance targets.

The contents of the Roadmap v4 describe the vision for the Geospatial Platform and its offering of data, services, applications, and infrastructure. Concepts and approaches put forth in the Roadmap will be further clarified and refined by supplemental documents that may include, but are not limited to, implementation plans, business plans, technical documents, project plans, and best practice examples.

4. INTRODUCTION

"In 2010 and 2011, Federal data managers for geospatial data will move to a portfolio management approach, creating a Geospatial Platform to support GeoOneStop, place-based initiatives, and other potential future programs.
This transformation will be facilitated by improving the governance framework to address the requirements of State, local and tribal agencies, Administration policy, and agency mission objectives.
Investments will be prioritized based on business needs.
The Geospatial Platform will explore opportunities for increased collaboration with Data.gov, with an emphasis on reuse of architectural standards and technology, ultimately increasing access to geospatial data."

FY 2011 President's Budget[3]

The Geospatial Platform will contribute to finding solutions to information management challenges of the Nation by organizing and sharing geospatial assets needed to solve the problems of the 21st century. In today's world, problems are multifaceted and complex, and require diverse information and collaboration to solve. One common feature of issues and events is that they occur in places and often impact surrounding neighborhoods, cities, regions, or global communities. Geospatial tools and the maps they display, widely popularized through Google Maps, Microsoft Bing, and personal global positioning systems (GPSs), can serve as the basis for understanding complex issues and relationships among multiple sources of information. Geospatial tools do much more than place locations on a map. These tools, working in concert with quality geospatial and tabular data of all kinds, provide an unparalleled ability to synthesize information, visualize patterns, and clearly communicate complex ideas. Geospatial tools greatly enhance data sharing, decision making, and functional collaboration across organizational and programmatic information silos. Information visually portrayed or communicated through maps promotes understanding and facilitate collaboration among decision makers, their partners, and the public. These unique features make geospatial assets among the most important strategic information assets for the Nation.

Governments at all levels and private sector businesses use geospatial technology to meet a wide range of programmatic, decision making, and business needs. Federal agencies and their State, regional, local and Tribal partners spend hundreds of millions of dollars on geospatial assets. Many governments garner the most value from their investments by finding ways to reuse or repurpose geospatial assets and reduce redundancy through coordination with partners. Two examples of existing cross-government geospatial cooperation include:

- The Department of Homeland Security, the National Geospatial-Intelligence Agency, the US Geological Survey (USGS), and the Federal Emergency Management Agency leverage their collective buying power to purchase commercial imagery that enable them to protect the Homeland and the Nation's precious natural resources at a reduced cost.

- To efficiently create an accurate National Hydrography Dataset (NHD), USGS, the Environmental Protection Agency, the US Forest Service, and the Bureau of Land Management coordinate efforts to avoid duplicative data collection. These agencies also foster a State stewardship program to facilitate the inclusion of detailed data collected by State governments into the NHD. This coordination supports higher quality data and reduces costs.

Examples of Government Geospatial Coordination

- The National Map
- Mortgage foreclosure tracking
- The Multi-Resolution Land Cover Consortium
- The Geospatial Line of Business
- Homeland Infrastructure Foundation-Level Data Working Group
- Imagery for the Nation
- Digital Coast

The Geospatial Platform will encourage coordination and strategically manage existing geospatial assets as a portfolio to maximize their value and increase their potential to solve problems and support everyday business needs in a manner that increases accountability and transparency.

As directed in the President's FY 2011 Budget language, the Geospatial Platform will advance geospatial asset management and access to geospatial data, services, and applications by:

- Establishing a portfolio management approach for geospatial assets that promotes sound management practices and prioritizes cross-government investments based on business needs and return-on-investment (ROI) to improve the overall quality of geospatial assets.
- Creating a functional, standards-based approach to deliver geospatial assets that become the foundation on which mission critical components are built once, then shared to support place-based and other initiatives.
- Expanding the geospatial governance framework to address the requirements of State, regional, local and Tribal agencies, Administration policies and Federal agency mission objectives.
- Building off the successes of Data.gov and Geodata.gov to promote reuse of architectural standards and technology.

4.1. Purpose of this Document

The purpose of this document is to serve as a roadmap for development of the Geospatial Platform. This document:

- Introduces the Geospatial Platform.
- Establishes a vision for the Geospatial Platform.
- Describes the key elements of the Geospatial Platform.
- Presents an initial vision for a Business Model (Section 5), which describes the business institution, potential customers, and a financial approach.

4.2. Vision of the Geospatial Platform

The vision for a fully implemented Geospatial Platform is that customers have access to geospatial assets including data, services, applications, and infrastructure that meet their business needs.

For government customers, it is envisioned that the Geospatial Platform offering will support reuse and multi-purposing of existing assets, economies of scale, and implementation of standards to meet mission goals and implement business processes. Governments at all levels will be able to better manage resources for efficiency and improve and validate assets through repeated use, stakeholder collaboration, and customer feedback. The Platform will serve as a vehicle to leverage the geospatial expertise and assets within individual organizations to meet the needs of customers resulting in improved efficiency and programmatic outcomes, increased savings, and enhanced geospatial capacity and utilization. Through efficient management of geospatial assets, the government will serve its citizenry with the capability to access, visualize, and analyze information and performance measures needed for a transparent government.

4.3. Purpose of the Geospatial Platform

The purpose of the Geospatial Platform is to:

- Provide the technology and organizational framework necessary to support missions and Administration priorities (e.g., Data.gov and Place-Based Policy initiatives) with timely and accurate geospatial data and spatial analytical capabilities.
- Promote sustainable governance and collaboration by promoting sharing of geospatial assets and engaging and supporting communication across stakeholders.
- Enhance interoperability.
- Ensure accountability and transparency of geospatial and related investments through portfolio management.
- Attain cost-savings and economies of scale through collaborative acquisition, development, and management of geospatial assets.
- Support and implement standards and processes that establish data quality and information assurance as criteria for assets included in the Geospatial Platform portfolio.

4.4. Definition of the Geospatial Platform

The Geospatial Platform will offer access to a suite of geospatial assets including data, services, applications, and infrastructure that will be known as the geospatial Platform offering within this document.

The Geospatial Platform will support an operational environment, www.GeoPlatform.gov, where customers can discover, access, and use shared data, services, applications, and when appropriate, infrastructure assets.

The Geospatial Platform is underpinned by:

- A segment architecture, aligned with the Federal Enterprise Architecture (FEA[4]), that emphasizes reuse of open and interoperable standards and technology, and supports increased access to geospatial assets. The Geospatial Platform will aim for a service-oriented architecture based upon common, secure, and scalable open-standards based technologies and may include cloud computing.
- Collaborative investment and portfolio management processes that enable partners and customers to leverage resources and share the costs of shared geospatial services.
- Policies and governance structures to ensure sound management practices and effective partnerships that address the requirements of Federal, State, regional, local, and Tribal organizations, Administration policy, and agency missions.
- A Managing Partner that services as a government focal point responsible and accountable for coordination and provision of data and services provided by the Geospatial Platform.

Geospatial Platform assets will be managed as a portfolio driving toward a scenario where the following characteristics are present:

- High quality and timely geospatial data, services, and applications are easy to discover and obtain.
- Customer needs are identified, planned, budgeted, and met in a geospatial context.
- Long-term costs of geospatial asset development, delivery, and access are reduced, duplicative efforts are minimized, and new business markets are developed.
- Effective commercial off-the-shelf systems and contractual business support operations are acquired more efficiently to replace legacy geospatial applications.
- Partners leverage non-geospatial assets with the goal of integrated information sharing.
- Collaborative management of geospatial assets occurs across all levels of government.
- Adaptable, proactive, and inclusive interactions are promoted among all stakeholders.

4.5. A Culture of Transformation: Changing the Game

Over the past decade, stakeholders in the geospatial community have witnessed the mixed success of numerous Federal initiatives designed to help improve, coordinate development, and manage geospatial assets. An important question now being asked is —What is different about the Geospatial Platform?" The following transformations differentiate the Platform from past efforts and will help fulfill the vision of the Geospatial Platform.

Toward Shared Leadership

A key strategy will be the ongoing outreach, consultation, and shared decision making with external stakeholders to develop a shared vision and shape and implement the Geospatial Platform.

The Roadmap identifies a variety of roles for stakeholders in the Geospatial Platform initiative.

The Managing Partner is committed to a collaborative governance approach. Stakeholders will be engaged continuously through in-person meetings, teleconferences, and social media (i.e., —Government 2.0" tools) to support the building of relationships, sharing of assets, as well as development and implementation of the Platform.

The Geospatial Platform will educate other communities about the power of geospatial tools and engage other, new communities of interest. Therefore, the Managing Partner is also committed to including Chief Information Officers (CIOs), Chief Financial Officers, government leadership, and elected officials in Geospatial Platform planning and implementation.

Toward a Business Perspective and Common Management Practices

Governments at all levels have an opportunity, through the Geospatial Platform, to manage the operations and delivery of geospatial assets using business-like practices.

A business approach described in a Business Model (Section 5) will lead to:

- A focal point for accountability and management of geospatial assets (Managing Partner).
- Identification of customer requirements and a clear definition of the Geospatial Platform offering.
- A plan to finance the operations of the Geospatial Platform.

Closely aligned with the development of the Business Model will be implementation of consistent management processes across the Geospatial Platform. These processes will include:

- *Portfolio Management* of geospatial assets where investment priorities are set based on customer business requirements.
- *Project Management* where each Platform implementation phase is managed as a project, utilizing industry accepted project management processes and techniques.
- *Partner and Customer Relationship Management* where partners will be treated as business partners with clearly defined and agreed upon expectations, requirements, and benefits. Customers will have the opportunity to express their needs and shape the offering. Both partners and customers will be encouraged to evaluate and offer feedback on the scope and performance of the Geospatial Platform.

Toward Sufficient and Stable Resources

The Geospatial Platform strategy includes moving away from managing and coordinating development of geospatial assets on a volunteer, ad-hoc, or —other duties as assigned basis". The Managing Partner will aim toward financial sustainability by identifying dedicated staff, funding, and revenue streams, and focusing on developing an offering that maximizes user benefit, cost-avoidance, and ROI. The Business Model begins the process of identifying the costs and funding/revenue streams associated with the Geospatial Platform.

Toward Open Government, Transparency and Accountability

Geospatial technologies provide an effective means to increase transparency and accountability not only by visually presenting information, but also by improving the ability to engage citizens, evaluate alternatives, and understand complex interactions. Implementation of the Geospatial Platform readily supports key Administration Open Government initiatives that emphasize government-to-citizen communication, accountability, and transparency including Data.gov, Apps.gov, Recovery.gov, and Place-Based Policy initiatives. Geospatial tools can be integrated into an electronic-government (E-Gov) environment, providing citizens with around-the-clock access to information about governmental activities, and facilitating opportunities to provide feedback and two-way communication. The Geospatial Platform will also deliver online services to provide access to information via interactive maps that allow direct access to useful government data. These interactive maps can aggregate information from disparate sources, facilitating rapid analysis for users. Maps, visual displays, and geo-enabled[5] dashboards can play valuable roles in monitoring, evaluating, tracking, reporting, and supporting decision making.

Toward Addressing Administration Priorities

Under the umbrella of Federal IT directives, the President‘s FY 2011 Budget[6] identifies a number of IT funding priorities, including the Geospatial Platform. Many goals of the Geospatial Platform are complementary to the goals of IT priorities: to reduce fragmentation, to streamline operations, transform customer service, and maximize the ROI from IT. The Geospatial Platform will contribute to attaining these goals by aligning with several Administration IT priorities in the following ways:

- The Platform will be linked to *Data.gov* and *Geodata.gov* to provide a source for shareable, reusable, and trusted geospatial data.
- Establishing a portfolio management approach for geospatial data, services, applications, and infrastructure as called for in the FY 2011 Budget provides a means of effectively selecting and tracking investments in geospatial assets as suggested in —*Managing the Federal IT Portfolio."*
- The new *FEA* will follow an improved model of development that will, *"...redesign IT* in key business areas from the ground up, based on central Federal platforms... [and will] provide an *interoperable, secure, and cost-effective* Federal IT *enterprise."*

- *Cloud computing*, web-based data, services, and applications, and infrastructure made available through the Geospatial Platform will support efforts to *consolidate Federal data centers* or help reduce the need for new data centers.
- The Platform can help eliminate fragmented purchasing of IT and *leverage the Federal Government's buying power* through Enterprise License Agreements (ELAs), Service Level Agreements, and other vehicles that promote sharing of and access to reusable geospatial assets.
- The Geospatial Platform will be focused on *customer service* and finding solutions to large and small problems that affect the Nation. *Transparency, accountability, and interoperability* are keys to effective decision-making, clear communication, and timely response.
- *Broadband Access for Americans* will be incorporated into the Platform.
- Information sharing networks must be *secure* and *protect privacy.* Some of the recommendations laid out in the Shared Infrastructure Pillar (Appendix B) reflect these ideas, specifically relating to the —Identity, Credential and Access Management (ICAM) Roadmap and Implementation Guidance."

Toward Matters and Places of National Importance

In August 2009, the White House instructed the heads of Federal agencies to begin developing effective Place-Based Policy initiatives for the FY 2011 Budget. The Administration has identified the use of place-based planning as a priority in its efforts to guide economic development, preserve natural resources, minimize government duplication or contradictory policies, and use geospatial technology for the benefit of the Nation. A focus on metropolitan areas, ecosystems, or regions can create synergies across government programs that allow agencies to share information and coordinate in ways not previously thought possible. Emerging national issues, such as climate change, health care, or childhood obesity can also be understood and managed using place-based approaches. As geospatial solutions are developed to address key issues, the Geospatial Platform can be used to share and leverage technical approaches, solutions, and lessons learned.

While much of our society and infrastructure is supported by available geospatial assets, other issues and places, ranging from rural and coastal communities to natural hazards and homeland security, need more attention. The Geospatial Platform will serve as the focal point to tap the potential of issues, places and users.

Place-based Policies

"Place-based policies leverage investments by focusing resources in targeted places and drawing on the compounding effect of well-coordinated action. Effective place-based policies can influence how rural and metropolitan areas develop, how well they function as places to live, work, operate a business, preserve heritage, and more. Such policies can also streamline otherwise redundant and disconnected programs" -From Office of Management and Budget (OMB) Memorandum 09-28

4.6. Challenges and Keys to Success

The time is ripe for a successful Geospatial Platform:

- The President's FY 2011 Budget identifies the development of a Geospatial Platform as a priority IT initiative for Federal agencies. This recognition engages the Office of Management and Budget to develop a plan (the Roadmap) under the auspices of the Department of the Interior (DOI).
- Technology and standards exist to develop and deliver the vision of the Geospatial Platform. Infrastructure, including the Internet and cloud computing, has matured to support the sharing of geospatial assets across multiple organizations. There is a wealth of robust geospatial assets being developed across all levels of government and the private sector. Industry IT and open and interoperable geospatial standards are available to allow sharing and reuse of assets.
- Past geospatial coordination efforts, including NGAC and the National States Geospatial Information Council (NSGIC) efforts, GOS, the Geospatial Line of Business (Geo LoB), and the Digital Coast have developed a clear path that enables us to move forward. The Platform is the logical next step in the evolution from dispersed governmental, private, and academic geospatial assets into a cohesive NSDI.

Spurred by this evolution and innovation and an ever-growing acknowledgement of the value of geospatial information, the Platform represents a valuable contribution as a component of the NSDI. Still, the Geospatial Platform will face challenges similar to those experienced by other initiatives.

Challenges

The success of the Geospatial Platform will depend on strong and dedicated leadership that can overcome a number of challenges associated with the development of the enterprise. Likely challenges and potential mitigation strategies are described below.

Identifying Requirements

There are many different, and sometimes conflicting, requirements among customers. One challenge will be clearly identifying and documenting various business requirements so that the Geospatial Platform offering meets those requirements. The Managing Partner will put forth maximum effort to connect with stakeholders and potential customers to identify requirements, and to ensure that their requirements are met through the Platform offering. Using automated tools, the Managing Partner will conduct initial and ongoing market surveys and needs assessments to understand what offerings are most desired by customers. The Managing Partner will remain nimble to assess and meet changing or evolving customer requirements over time.

The Managing Partner is committed to gathering and staying current on existing and evolving requirements of non-Federal governments, while appreciating the constraints resulting from differences in fiscal years and budget processes.

Evolution of Technology

As technology advances, customer requirements will quickly expand and change. The Managing Partner will quickly react to changing requirements and adapt to, utilize, and offer new tools.

Identifying Solutions

The Managing Partner may find it challenging to identify assets or solutions that meet customer needs. The Managing Partner will conduct a capabilities assessment of potential partners to understand potential assets to meet customer requirements and will work to engage and incentivize partners to share assets through GeoPlatform.gov. Once an offering is available through the Geospatial Platform, the Managing Partner will work to ensure that partners continue to manage their data, services, and applications, innovate, and remain actively engaged with the Geospatial Platform to continue meeting customer requirements.

Licensing and Sharing Constraints

Licensing and sharing constraints may exist in some government and private organizations, creating difficulties to leverage and share geospatial assets. The Managing Partner is committed to working through these challenges to promote sharing and efficiency.

Leadership

Leadership should not only come from within the Managing Partner, but also from agency political leadership in order to provide the full scope and level of support Platform activities. Some agencies may be reluctant to adopt new policies, procedures and funding plans. In order for the Platform to achieve tangible results, leadership must receive stakeholder buy-in by demonstrating benefits (e.g., ROI, optimization of resources, etc.) and providing incentives for participation (e.g., marketing, hosting, etc.). Finally, at all times, the Platform leadership must focus on the measurable results of ongoing activities to maintain user and stakeholder trust in the effort.

Financial Sustainability

Another crucial component for determining Geospatial Platform success is the long-term financial strategy. Unless the Managing Partner determines a sustainable long-term financial approach, the Geospatial Platform offering may not be able to meet stakeholder expectations. If the incentives and/or offering are not compelling, stakeholders may not be willing to invest in the Geospatial Platform activities and the Geospatial Platform may prove to be unviable in the long-term.

Keys to Success

The Geospatial Platform is a highly ambitious yet achievable initiative. Realization of the vision is largely within the capacity of the geospatial community. In order to achieve the vision of the Geospatial Platform, the community must:

- Act as one to develop, articulate, and commit to a shared vision.
- Set aside past differences and work together toward common goals.

- Develop a business plan that includes strategies and incentives for sharing and partnerships.
- Establish a governance approach that provides for meaningful input and consultation with all stakeholders.
- Initiate outreach and communications using both traditional and new media to include and educate stakeholders.

This initiative will also require upfront and ongoing support of the Administration including:

- Designating and empowering the Managing Partner.
- Providing political leadership to support the FGDC enterprise.
- Providing resources:
  - For hosting content as a primary incentive for participation by partners.
  - Long-term financial support through appropriations starting in FY 2012 for infrastructure, new offerings, and Platform management.
- Creating a supportive policy environment that encourages agency participation (e.g., finalization and enforcement of A-16 Supplemental Guidance, geospatial segment architecture, grants and contracts language).
- Developing and maintaining structures for accountability. Developing a strong organizational relationship between the Geospatial Platform and other priority IT initiatives including Data.gov.

5. PILLARS OF THE GEOSPATIAL PLATFORM

In June 2010, the Geospatial Platform was envisioned (via the Roadmap) as five inter-related components (Pillars) are depicted in a conceptual model shown in Figure 1.

The organization of the Platform can be logically grouped into two discrete categories:

Offering – The suite of assets delivered through the Geospatial Platform that comprises the Geospatial Platform Portfolio:

- Data, Services, and Applications.
- Infrastructure.

Supporting Processes – The business controls necessary to ensure the Geospatial Platform offering is delivered:

- Segment Architecture.
- Governance.
- Portfolio Management.

Appendices A-B provide details of each Pillar as envisioned in June 2010. The March 2011 working definitions of the Platform components are provided below.

5.1. Pillars Defined

Offering

The Geospatial Platform offering is defined as the suite of geospatial assets delivered to customers through GeoPlatform.gov, including geospatial data, services, applications, and infrastructure. When viewed together, this suite of assets can be viewed and managed as a set of inter-related portfolios.

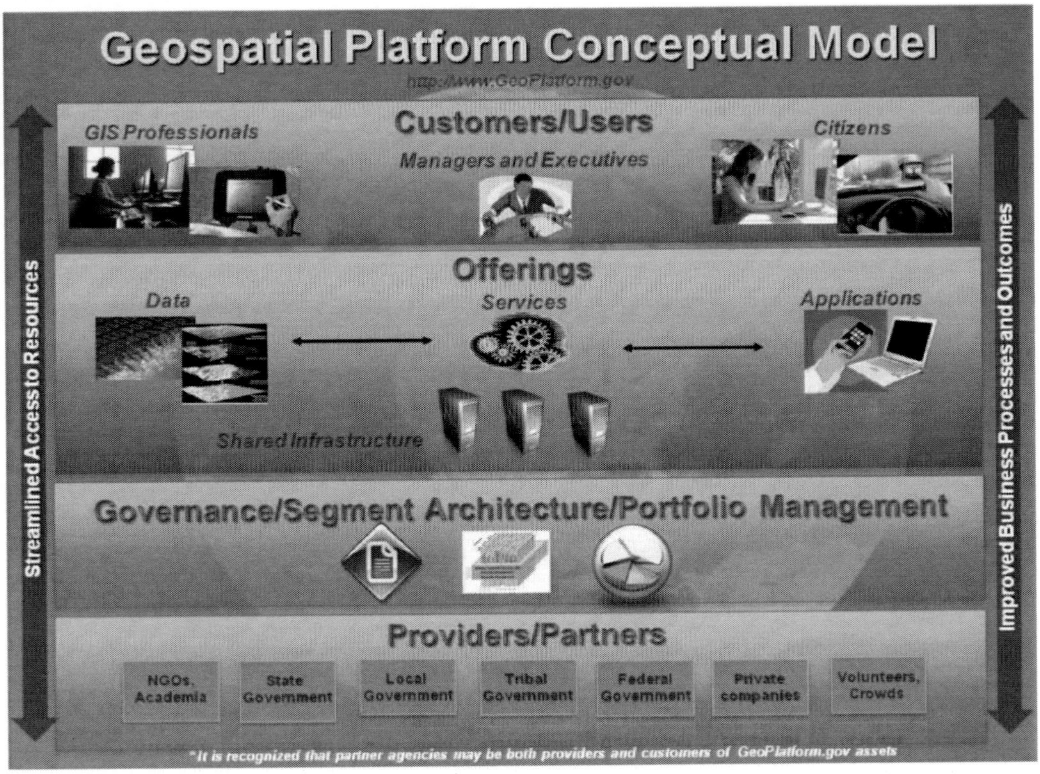

Figure 1. Geospatial Platform Conceptual Model.

Data: includes individual datasets, integrated data (such as base maps), or other products derived from multiple datasets. These assets will provide foundational geospatial data that can be trusted, used reliably, and shared across organizations. Data can be provided by governments at all levels or non-governmental organizations.

The Geospatial Platform web based services will help customers discover desired geospatial data, through an interface with Data.gov or other similar data portals/registries.

While all data will be discoverable, the Geospatial Platform will promote only data that meet the Platform portfolio inclusion criteria (to be developed). Data that are included in the portfolio will be —tagged as trusted Platform assets. The Platform portfolio data, and all Platform assets, will be managed using life cycle management practices.

Services: provide a consistent result to all users (e.g., geocoding, metadata editor). The Geospatial Platform will offer access to services that can be used by multiple agencies as stand-alone capabilities or as building blocks to develop applications. Services offered to customers will include technical geospatial services as well as support services (e.g., acquisition, requirements development).

Applications: consist of a set of tools or capabilities that enable a user to select how they utilize the tools and capabilities to perform geospatial visualization or analysis to achieve their results. Applications may leverage one to many different services to conduct analysis and return a result to the user. The Geospatial Platform will offer access to applications that can be downloaded, customized, and used to meet customer business needs.

Infrastructure: includes both physical and logical components of IT that can be leveraged by multiple customers. The physical shared infrastructure includes (but is not limited to):

- Data centers and data repositories, including the cloud.
- Networks, including Internet and intranets.
- Host platforms for data, services, and applications, including portals and services.

The logical shared infrastructure includes (but is not limited to):

- Network routing services.
- Access control and security.

Supporting Processes

Segment Architecture: provides a process-driven approach to designing and deploying large computing components in the context of a broader enterprise, which results in a design that can be readily deployed in an actionable solution architecture by partners that collaborate on geospatial data and services.

Governance: Processes by which parties with a stake in the Geospatial Platform are afforded an opportunity to shape its structure, functions and capabilities. The ultimate vision is to be a national system with shared governance.

Portfolio Management: establishes a set of processes by which the Geospatial Platform will manage geospatial assets contained within the offering. The Geospatial Platform will strategically manage existing geospatial assets as a portfolio to maximize their value and increase their potential to solve problems and support everyday business needs in a manner that increases accountability and transparency.

For Federal agency partners, portfolio management pertains to the management of investments in geospatial data, services, applications, and infrastructure[7] assets. This includes setting priorities and selecting assets, and then allocating resources to asset development/acquisition, management, and evaluation to maximize enterprise value and obtain the best possible strategic impact of each investment.

For non-Federal partners, portfolio management is more limited to such areas as defining criteria for inclusion of assets on the Platform, brokering partnership agreements, and customer relationship management.

6. Business Model

As described in Section 4, the Geospatial Platform will offer access to geospatial assets through GeoPlatform.gov. The Platform will function according to a Business Model in order to maximize value to customers and minimize expenses to ensure that customer requirements are met cost-effectively.

The Geospatial Platform Business Model includes the following components:

- Business Institution – the organizational elements and processes that support the Geospatial Platform.
- Customers – the consumers of the Geospatial Platform offerings.
- Financial Approach – the anticipated costs of and potential approaches for financing implementation and ongoing management of the Platform.

The business components of Geospatial Platform are portrayed in Figure 2.

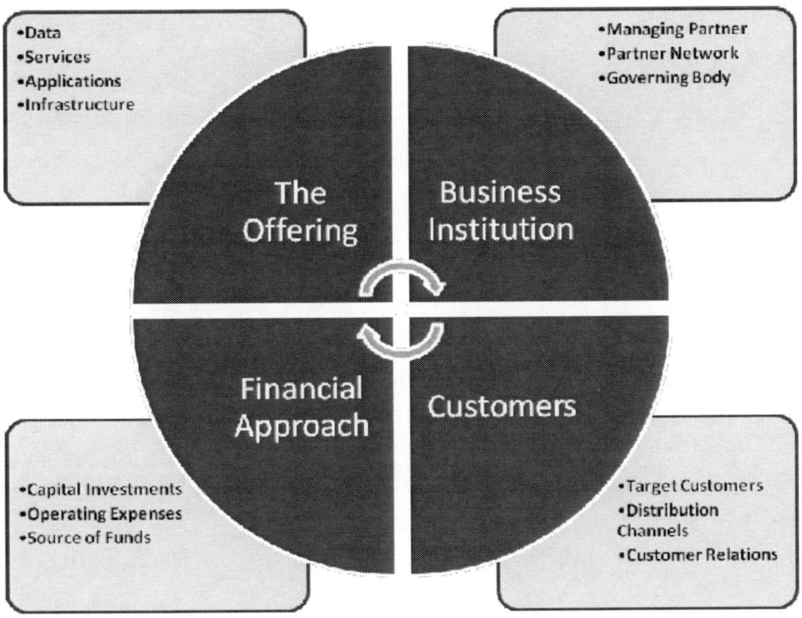

Figure 2. Geospatial Platform Business Componenets.

6.1. The Business Institution

The Geospatial Platform Business Institution will contain related elements that include:

- *Managing Partner* that will coordinate the Platform's activities and develop, deploy, operate, and manage the offering.
- *Partner Network* that will be comprised of providers of data, services, applications, and infrastructure and coordinated by the Managing Partner.

- *Governing Organization* that will be comprised of partners, customers, and other stakeholders from all levels of government and non-governmental organizations to ensure that stakeholder voices are heard and that customer requirements are met.

The Business Institution for the Geospatial Platform can be viewed in Figure 3 as a network of providers (partners) and business processes that are brokered and managed through the Managing Partner.

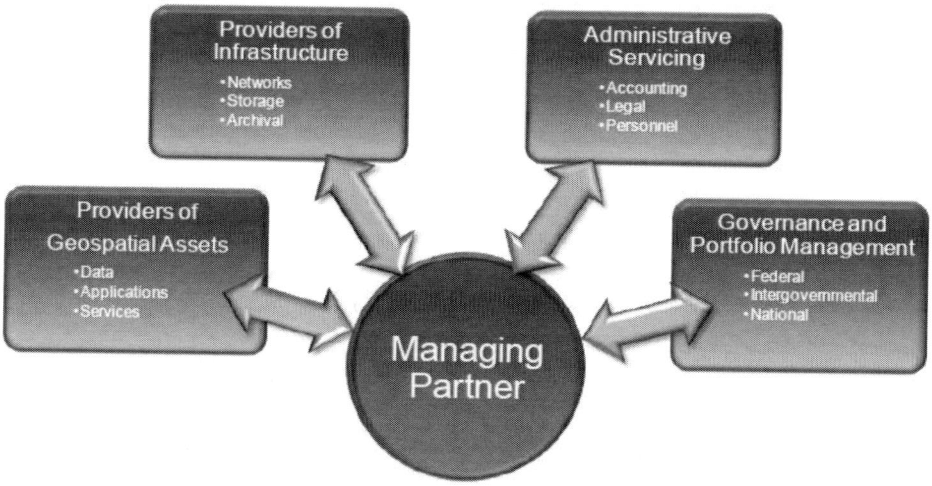

Figure 3. The Geospatial Platform Business Institution.

Managing Partner

The Managing Partner will coordinate the Platform's activities and develop, deploy, operate, and manage the offering. The Managing Partner needs full-time resources to carry out its functions. Important functions of Managing Partner will include, but are not limited to, establishing and managing relationships with partners; facilitating requirements analyses and developing the offering; managing the portfolio; customer relationship management; and outreach and marketing efforts.

The Managing Partner will serve as the broker and manager of the Partner Network and the primary interface with customers.

Until a Managing Partner is selected, the role will be met in the following way:

- DOI will serve as the interim Managing Partner.
- The Chair of the FGDC (or designee) will serve as interim Executive Sponsor of the Geospatial Platform.
- An interim Project Manager will be donated by one of the Federal agencies.
- The FGDC Secretariat and Federal member agencies will provide interim staff support to the Geospatial Platform.

A process for selection and oversight of the longer-term Managing Partner will be developed in consultation with OMB and the FGDC Chair and Steering Committee. Initial criteria for a successful Managing Partner should include:

- Ability to negotiate and manage formal partner relationships (e.g., contractual agreements, service level agreements, etc.).
- Capability to conduct geospatial requirements analyses.
- Credibility with the geospatial stakeholder community.
- Portfolio management capabilities.
- Customer relationship management capabilities.

Partner Network

Partners provide content (i.e., geospatial assets including data, services, applications, and infrastructure) to the Geospatial Platform. They may be governmental or non-governmental organizations.

Initially the Managing Partner will focus on identifying customer requirements and identifying organizations with assets available to meet those requirements.

The Managing Partner will reach out to these organizations to encourage them to become a member of the Partner Network and share their assets through the Geospatial Platform.

The Managing Partner understands that some organizations have geospatial assets that can be quickly shared with few added resources to the partner, while other desirable assets will require more resources. Explicit and achievable incentives will be defined to help encourage partnerships that might include, but are not limited to hosting capacity, technical support, and exposure of partner services to the broader community, or reimbursement for services.

For instance, State organizations might find that database-hosting services offered by the Platform are a valuable incentive to provide services. In essence, States may be willing to manage and share key geospatial data sets if the Platform provides hosting services in return. Also, for some private sector organizations exposure of their value added applications might serve as an incentive to offer services to the Platform.

When assets are shared with the Geospatial Platform, they will become part of the Geospatial Platform portfolio and benefit from the Geospatial Platform portfolio management processes and marketing.

Once providers become partners, they will be required to meet —rules of engagement" included in negotiated contractual, license, or service level agreements, to continue their partner status. These —rules of engagement" and business processes will be clearly defined and monitored for compliance and performance reporting.

Governing Organization

The Managing Partner is committed to managing Platform operations through a collaborative approach that includes all stakeholders. The Geospatial Platform will develop a shared governance model that involves non-Federal entities. The President‘s Budget direction notes that the Platform will be —facilitated by improving the governance framework to address the requirements of State, regional, local, and Tribal agencies, Administration policy, and agency mission objectives.‖ As a first step, the FGDC supported the establishment of an NGAC Intergovernmental Subcommittee to provide ongoing advice and feedback to the Managing Partner. The NGAC Intergovernmental Subcommittee includes representatives from a broad cross-section of the geospatial community, including members from governmental, private sector, non-profit, and academic organizations.The Managing Partner will develop a process for long-term stakeholder engagement and shared governance of the Geospatial Platform in collaboration with stakeholders and in consultation with OMB.

6.2. The Customers

The customer drives business requirements and Platform offerings. Key customer segments may include:

- Federal Government.
- State, regional, local, and Tribal governments.
- Private sector.
- Academia.
- Non-profits.
- Citizens.

Customers can also be categorized as:

- Geographic Information Systems (GIS) professionals.
- Executives.
- Managers.
- Scientists.
- The general public.

During early implementation, the Geospatial Platform will focus on government customers.

The Managing Partner will act as the primary interface between the Geospatial Platform and its customers. The Managing Partner will conduct a market survey /needs assessment to better understand the target market and identify customer groups. These interactions will also help the Managing Partner to develop the processes of ongoing customer interactions and feedback.

The anticipated major categories of customer interactions are depicted in Figure 4.

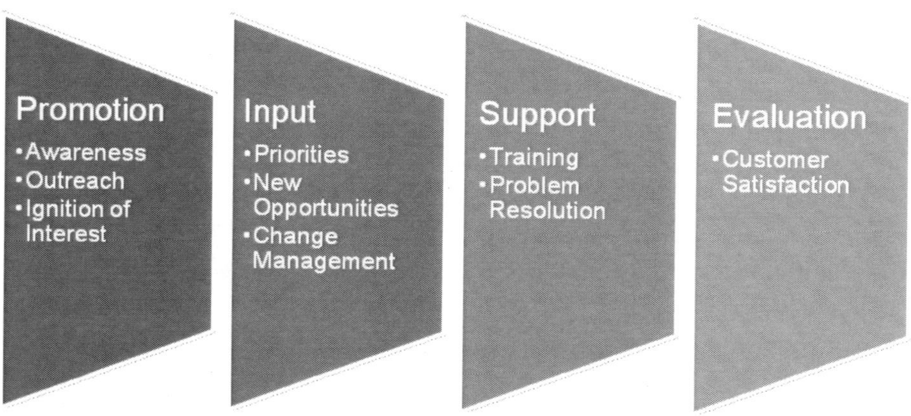

Figure 4. Major Categories of Customer Interactions.

6.3. The Financial Approach

The financial approach for the Geospatial Platform considers both the costs to develop and manage the offering and the sources of funds and other resources that will be used to cover the costs.

Costs/Expenses

Start- Up Costs

In the first stages of the Geospatial Platform development, capital investments will be necessary to ensure smooth implementation. Capital investments will include infrastructure, processing capabilities, and asset migration. The Geospatial Platform will utilize the resources of similar initiatives, such as cloud computing, data center consolidation led by the Federal CIO Council, and Electronic Capital Planning and Investment Control System.

The Geospatial Platform will leverage existing assets. Sharing the tools and technology developed by other agencies or partners not only reduces costs, but also speeds implementation of the Platform and builds upon already established best practices.

Over time, however, these investments will reduce the government-wide cost of geospatial technology by decreasing redundancy, increasing efficiency, and improving the tools and data available for executing agencies‘ core business processes.

Business Institution Costs

The Business Institution is the component of the Platform that creates expenses; these include both direct costs and overhead costs. Platform expenses, grouped by the core elements of the Business Institution, include:

Managing Partner

- Salary and benefits for Program Manager.
- Staff and/or contractual services for:
  - Customer relations and support.
  - Partner relations.
  - Geospatial technical oversight and management.
  - Administrative and financial management.
  - Marketing (e.g., promotion, pricing, product development, and research).
  - Outreach and communications.
  - Government accountability (e.g., budgeting, performance management, reporting).
- Office hardware and software to support Program Manager and staff.

Partner Network

- Provisioning of technology infrastructure for hosting.
- Acquisition of content (data, services, and applications) from partners/providers.
- Technical support and training for partners.

Governing Organization

- Staff and/or contractual services for stakeholder input and portfolio management.

- Meetings.
- Communications.
- Portfolio management process development and application.

Geospatial Platform Source of Funds and Other Resources

The Geospatial Platform is considering using four major methods to cover the costs of operations. These include:

Common Good Funding.

- Direct appropriations and/or reinvestment of cost-savings.
- Across the board assessments (e.g., Geo LoB, GOS).
- Reimbursable agreements under the general authority of the Economy Act.

Incentive-based Partnerships

- Intergovernmental partnerships that involve a government entity providing no-cost data, services, applications, or infrastructure to the Platform in return for receiving commensurate benefits.
- Intergovernmental partnerships that involve government entities sharing the cost of data, services, applications, or infrastructure in return for receiving commensurate benefits.

No-Cost Contributions from Partners

- Partners providing content as a part of filling mission requirements to create public domain geospatial formation (e.g., National Wetlands Inventory, *The National Map*).
- In-kind contributions from partners who are willing to cover the full cost of a dataset, application or service because the marginal costs beyond meeting their mission obligations is minimal (e.g., Palanterra).
- Donations of data, services, applications, or infrastructure from government agencies or their partners.

Income from Customers of Services

- Metered services (e.g., geocoding costs charged per transaction).
- Subscription fees.

In summary, the Geospatial Platform Business Model places new emphasis on managing a Partner Network of providers to deliver valuable and trusted geospatial assets that meet the needs and requirements of key customers.

7. IMPLEMENTATION AND PERFORMANCE GOALS

The Geospatial Platform is building upon multiple ongoing and related efforts, including Geo LoB activities, *The National Map*, development of the segment architecture, collaboration with Administration activities (e.g., *Data.gov, Apps.gov,* etc.), and development of related programmatic initiatives (e.g., development of a business plan and budget request for the Imagery for the Nation initiative).

This section describes the strategic path forward for development and implementation of the Geospatial Platform and preliminary thinking on performance measures. The Platform implementation approach will be informed by several key principles:

7.1. Long-term Vision

This document provides a conceptual vision, to be refined based upon collaboration and consultation with Geospatial Platform stakeholders. The vision of the Geospatial Platform is consistent with the Geospatial Segment Architecture, builds upon the portfolio management practices outlined in the A-16 Supplemental Guidance and aligns with the Administration's goals for ensuring transparency, democratizing public-sector data, and driving innovation.

7.2. Phased Implementation

The Geospatial Platform will be developed in a phased approach, with a bias toward action that will provide immediate improvements and useful products and components for customers. This incremental development approach will provide a steady stream of valuable offerings for customers. Development of the Platform will be informed by ongoing and frequent interactions with stakeholders, and will work to incorporate new requirements, feedback, and recommendations at each phase.

7.3. Collaboration with Stakeholders

The Geospatial Platform effort will involve substantial outreach and consultation with external stakeholders to reach a shared vision and develop a collaborative governance approach.

The Geospatial Platform aims to provide valuable assets for use by all customers.

Over the coming months, the Managing Partner will engage with stakeholders and potential partners through a variety of means, including the following:

- Consultation with the NGAC.
- Social media approaches.
- Capabilities assessments.
- Targeted sessions with key stakeholder groups.
- Dialogue and discussion at industry and professional meetings and conferences.

7.4. Project Management

The Geospatial Platform will be managed using best practices of project management, including a time box approach to phased implementation. Time boxing is a project management technique often used in planning projects and in rapid application development software development processes, where an implementation schedule is divided into a number of separate time periods (i.e., time boxes), with each period having its own deliverables and deadlines.

The initial implementation of the Geospatial Platform focuses on incremental implementation of existing components and capabilities using dedicated teams organized around the time box goals.

7.5. Measuring Performance

Performance measures are an integral part of understanding how the Platform meets customer needs, identifies successes and gaps, and provides insight into ROI.

In addition, performance measures extend transparency and enhance accountability among both internal and external stakeholders.

The Managing Partner is committed to finding the right blend of performance measures that will provide a practical, yet holistic look at the Platform and its offerings.

Performance measures link overarching goals of the Platform to the day-to-day operations and resources (human, intellectual, financial, and organizational), and answer key questions about performance, outputs, and outcomes including:

- Are the right people involved and are those with responsibilities meeting those responsibilities?
- Are stakeholder needs clearly defined and efficiently addressed?
- Are investments showing returns and value? Are resources measured and maximized? Are there opportunities for partnerships or collaboration?
- Are the right assets (i.e., data, services, applications, infrastructure) available and of high quality? Are the appropriate selection criteria in place? Are there opportunities for collaboration or consolidation?
- Does the infrastructure function as intended? Is access available when needed and to the right people?
- Are priorities selected efficiently? Are priorities organized and addressed appropriately?
- Are best practices and lessons learned captured?
- Have reporting mechanisms (e.g., dashboards, maturity models, etc.) and timetables been established?

Financial Performance and Investments

Financial performance metrics provide indicators highlighting the health of the Platform and can show ROI for assets within the Platform Portfolio (i.e., data, services, applications, infrastructure) as well as investments made in human capital that allow the Platform to address stakeholder needs.

- Cost avoidance in dollars from ELAs.
- ROI calculations for data investment.

- Cost-benefit analyses for services.
- Investment collaboration reviews for data, services, applications, and infrastructure or to set Platform-wide investment priorities.
- Opportunity cost analysis of partnerships.

Operational Performance Metrics

Operational performance metrics highlight the day-to-day operations and maintenance goals of the Platform and provide information on what data, services, or applications are available on the Platform as well as when, and what people are seeking/using within the Platform. Some examples of operational performance metrics might include:

- Number of partnerships established or partnership opportunities identified.
- Quantity of data, services, and applications available through the Platform.
- Number of downloads/site visits.
- Percentage of users who are repeat customers.
- Number of Federal, State, regional, local, or Tribal users.
- Average server or request response time.
- Percentage of time the Platform is online and available to users.

Customer Service Performance Metrics

Similar to operational performance metrics, customer service performance metrics highlight day-to-day Platform usage; however, unlike operational metrics above, these metrics focus on who is using the Platform and whether the Platform is meeting stakeholder business needs. Examples of customer service performance metrics might include:

- Number of customers.
- Number of downloads/website hits by user group (e.g., Federal, State, regional, local or Tribal government, private sector, academic, or citizen users).
- Ease-of-use or customer satisfaction survey scores.
- Number of listening/feedback sessions or number of comments received.

Governance and Compliance Metrics

Governance and compliance metrics relate to the degree to which Platform assets meet requirements and standards issued by authoritative bodies (e.g., International Organization of Standardization (ISO) Standards, FGDC metadata standards) or the degree to which stakeholders meet agreed upon roles and responsibilities (e.g., Platform stakeholder rules of engagement, leadership responsibilities, etc.). Some examples of governance and compliance metrics include:

- Percentage of data that are ISO/FGDC Standards compliant.
- Percentage of data with FGDC metadata compliance.
- Percentage of Federal data that are A-16 Supplemental Guidance compliant.
- Performance of each stakeholder in implementing roles and responsibilities.
- Participation by agency or participation by State.

- Percentage compliance with rules of engagement.

Portfolio Management Metrics

The Geospatial Platform portfolio will be measured and reported based on the following:

- Data maturity based on maturity model scores.
- IT dashboards.
- Investment review criteria and investment review scores.
- Implementation of roles and responsibilities specific to portfolio management (e.g., roles and responsibilities in the A-16 Supplemental Guidance for data portfolio management).

Platform Implementation Metrics

Geospatial Platform implementation will occur in a phased approach based on Platform priorities, funding, and partners' ability to implement standards, requirements, roles and responsibilities, and funding model. In order to track implementation progress, Platform implementation metrics will evaluate:

- Cost (baseline and variance).
- Schedule (baseline and variance).
- Scope (baseline and variance).

Performance against targets for metrics will be specific to each phase of implementation.

APPENDIX A. COMMON DATA, SERVICES AND APPLICATIONS, JUNE 2010

A.1. Definition

Common data, services and applications are recognized as official or trusted resources with a Federal lead agency steward (for data, "Dataset Manager" per A-16 Supplemental Guidance).

The Federal lead agency verifies and vouches for the quality and reliability of the resources, which may include commercial resources. (Examples include TIGER data, a geocoding service, and commercial GIS software.)

Common assets are managed as part of the Geospatial Platform portfolio and therefore must meet portfolio inclusion criteria (for data, see A-16 Supplemental Guidance).

Common Geospatial Data, Services and Applications are:

- Used by two or more entities to meet one or more geospatial
- business requirements (i.e., provide a common solution).
- Registered and discoverable through the Geospatial Platform portfolio.
- Provided in a standardized form, format, or process.

It is important to note that the term "common" as defined herein does not define the distinctions between authoritative, official, trusted, validated, or best available information.

This will be taken into account through the metadata, data/services/applications registry and user interface components of the Geospatial Platform and be determined by an evaluation of the source, content, quality and intended use.

***Common* Data, Services and Applications:**
A single solution, utilized by two or more agencies/partners, to meet: • A common requirement between agencies, or • Meet different requirements across agencies *One solution – multiple uses*

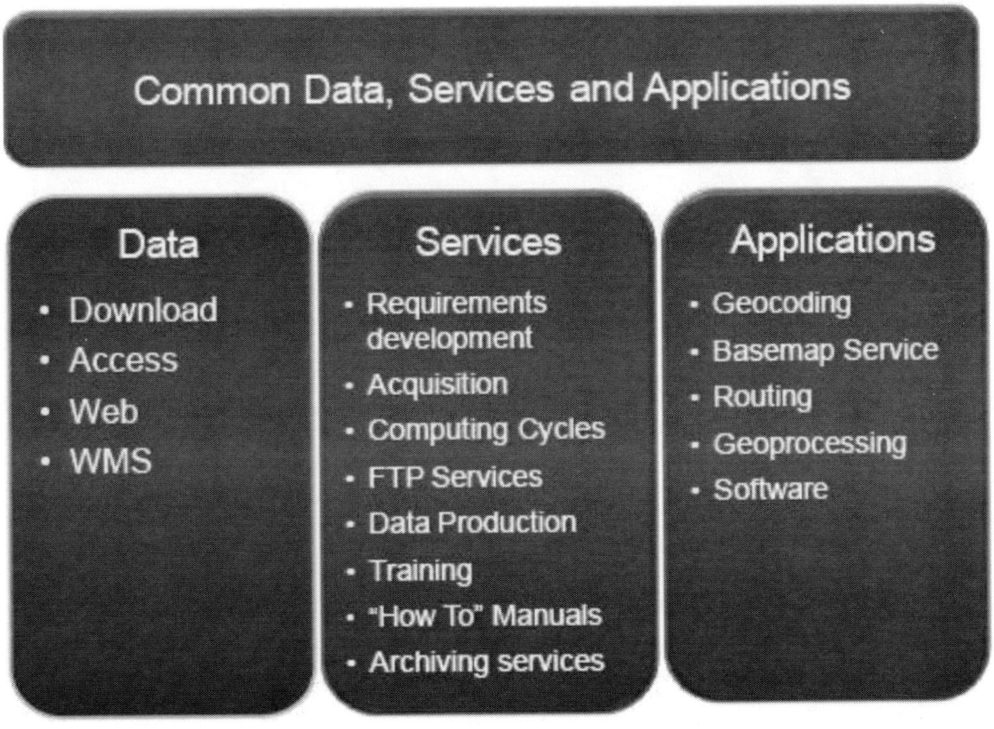

Figure A.1. Common Data, Services and Applications Offerings.

Common Data

The common data component of the Geospatial Platform portfolio includes any geospatial data selected by the Managing Partner because they meet inclusion criteria outlined in the A-16 Supplemental Guidance and one used by two or more Federal agencies and partners to meeting their business requirements. Common data for the Geospatial Platform will provide the foundation for meeting a broad spectrum of customer business requirements and are often created or acquired in response to a need.

Geospatial data are developed and managed to meet specific format, content and structural requirements that are based on the intended use. As new requirements are identified by the national geospatial community, or new data identified that meet defined requirements, the format, content, structure and even the number of datasets required may change. For example, a roads dataset may be used to perform vehicle routing by one agency and the same dataset used as a base map by another agency. In either case, there is a single dataset utilized by multiple agencies.

Common data include access to either original source data or data, which have been integrated to meet requirements for consistency. The Census TIGER/Line files are an example of a single formatted dataset that is used across multiple agencies to render demographic information. There are many types of access methods to geospatial data including online web services, data download and all other data access methods.

Common Services

The common services component of the Geospatial Platform portfolio includes all other types of geospatial services that provide a consistent capability to users (i.e., tools and services that support the execution and management of geospatial business processes and offline services). While peripheral to the analysis capabilities that form the core execution of the Geospatial Platform‘s function, these common services are key underlying and complimentary efforts that must be included as part of the overall Geospatial Platform portfolio.

Examples of the types of support services include requirements development and tracking tools, system computing cycles and file transfer protocol (FTP) services. Examples of offline administrative services include acquisition services (such as the Geospatial SmartBUY for geospatial software established through the Geo LoB), data production services and training. It is also envisioned that a hosting service will be one of the Geospatial Platform services offered for those agencies without the required capacity to host or that have security concerns about hosting all of the data that are required by other agencies, stakeholders and partners.

Common Applications

The common applications component of the Geospatial Platform portfolio includes any computer-based set of tools or capabilities that enables a user to perform geospatial visualization or analysis, including software, online applications and geo-processing required by two or more agencies/ partners/ customers.

This category of applications includes a broad spectrum of analytical support generated to facilitate the government‘s delivery of services to citizens. Applications may leverage data services to conduct analysis or computations and return a result to the user or they may provide the user interface between and with other services. The utilization of open standards will allow these various components to communicate and interoperate effectively.

A base map visualization service from rendered and cached geospatial data, a routing tool, a geocoding service and geographic information system (GIS) software are all examples of common applications. A potentially high-impact common application for the Geospatial Platform is a consistent, standards-based metadata editor. As all Federal agencies must comply with the FGDC metadata standard and create metadata for every geospatial dataset, it makes sense to develop one common metadata editor and deploy it across agencies. Agencies can avoid duplicative software development by using a common application that allows for easy compliance. If desired, agencies can implement a customized style for the metadata editor tailored to their individual needs.

Data.gov is an example of a successful common application through which Federal geospatial data are made available to the public. Data.gov was established recognizing that data are assets of the Federal Government and its citizens and provides public information for the public good. Efforts are underway to add additional geospatial functionality to the Data.gov infrastructure enabling users to view their selected data in a map.

A.2. Objective/Purpose

The objective of Common Data, Services and Applications is to ensure that data, services and applications are discoverable, accessible, usable and obtainable by all. The Geospatial Platform will expose and leverage existing resources, providing a means to identify and eliminate redundancies among agencies, and at the same time enabling a geospatial market by opening up these common services and requirements.

Common data, services and applications are achieved when:

- Standards are adopted and implemented by providers of data, services and applications.
- Publication of the data, services and applications follow a common protocol for registry, publishing and discovery.
- The full spectrum of data are published with fully compliant metadata removing any debate about the most authoritative/appropriate dataset.
- Common applications are developed using core requirements of multiple agencies upon which user interface and branding can be modified if necessary to better meet customer requirements.

A.2. Approach/Strategies

The Geospatial Platform will meet business needs of its customers, support transparency and access to information, and empower the Nation's citizens through access to a range of geospatial data, services and applications. Federal leadership must assist in lowering the barrier to publish data, services and applications by ensuring a common method and approach is adhered to by the community. The Geospatial Platform will implement an open government approach to standards based services publication and will adopt and enforce the use of open and interoperable standards to simplify discovery, access and dissemination of data, services and applications. The Geospatial Platform will then be populated with common data, services and applications. As the Geospatial Platform matures, a number of guidance documents and how-to manuals will be produced to encourage participation by all members of the national geospatial community in the effort.

To achieve the Geospatial Platform objectives, the processes and frameworks will be developed to:

- Endorse and enforce standards.
- Identify/inventory/catalog available geospatial data, services and applications.
- Identify common business requirements they meet or are utilized to support.
- Identify new or evolving common business requirements that can be supported by geospatial assets.
- Identify the functions, the services and applications performed.
- Identify those data, services and applications that are utilized by multiple agencies to meet common business needs.
- Establish the mechanisms to govern, operate, evaluate and service the common components.

A.4. Priorities

The Geospatial Platform priorities with respect to common data, services and applications include:

- Discovery of geospatial data, services and applications.
- Initial inventory and selection of geospatial data, services and applications (also see Portfolio Management objectives and priorities).
- Endorsement and enforcement of standards.
- Identify priorities for new common data, services and applications.

Discovery of Geospatial Data, Services and Applications

A fundamental goal of the Geospatial Platform is to enable providers and customers to contribute, access and leverage data, services and applications, and provide a clear method for discovering valuable resources. Common and discoverable data, services and applications are often the starting point for identifying data, services and applications that have broader value to multiple partners and may be included in the Geospatial Platform portfolio and delivered as common resources. One Geospatial Platform service will include evaluating the data, services and applications, and the requirements they are designed to fulfill.

While not all data, services and applications will be included in the Geospatial Platform portfolio, all Federal investments in geospatial data, services and applications should be discoverable and inventoried as part of the national strategy. The Geospatial Platform will encourage all Federal agencies and partners to share and leverage their geospatial assets.

Federal Executive branch agencies should publish metadata for all geospatial datasets and applications to *Data.gov*, the authoritative repository for Executive branch data. Another system service will be developed for non-Executive Federal and non-Federal data and applications as part of the Geospatial Platform.

These systems will work together to ensure that all geospatial data and applications are discoverable to customers.

An initial asset inventory and selection of geospatial data, services and applications for the Geospatial Platform portfolio, will need to occur in order for the Managing Partner to know what geospatial data and applications exist and how they support the business requirements of the Federal Government and its partners.

The registry developed using Data.gov and its counterpart (for non-Executive Federal and non-Federal data and applications) will serve as the initial government wide inventory of geospatial data, services and applications. This initial inventory activity will be conducted to create a baseline Geospatial Platform portfolio, which will be routinely maintained on at least an annual basis.

From this initial inventory, geospatial data, services and applications will be selected for inclusion in the Geospatial Platform portfolio.

Processes for data selection are outlined in the A-16 Supplemental Guidance, whereas similar processes for selecting services and applications will need to be developed (see Portfolio Management priorities).

Initial Inventory and Selection of Geospatial Data, Services and Applications

To select assets for the Geospatial Platform portfolio, the Managing Partner must know what geospatial data and applications exist and how they support the business requirements of the Federal Government and its partners. The registry developed using Data.gov and its counterpart (for non-Executive Federal and non-Federal data and applications) will serve as the initial government wide inventory of geospatial data, services and applications. This initial inventory activity will be conducted to create a national geospatial inventory, which will be routinely maintained on at least an annual basis.

From this initial inventory, geospatial data, services and applications will be selected for inclusion in the Geospatial Platform portfolio. Processes for data selection are outlined in the A-16 Supplemental Guidance, whereas similar processes for selecting services and applications will need to be developed (see Portfolio Management priorities).

Endorsement and enforcement of standards (also see Segment Architecture)

An open government approach to standards based services publication will fuel the demand for more geospatial data and robust application development as more and more location-based services are developed. Through the FGDC, the Geospatial Platform will adopt and enforce open and interoperable standards to leverage the work already accomplished by national and international standards organizations. Standards will simplify discovery, access and dissemination of data, services and applications through the Geospatial Platform. Shared standards that shall be formally evaluated for adoption by the Federal Government include, but are not limited to:

- Geospatial data publication standards as described by the Open Geospatial Consortium (OGC), the International Organization of Standardization (ISO), the International Committee for Information Technology Standards (INCITS) L1 committee that deals specifically with geospatial standards, and the FGDC including but not limited to the Web Services (features, map, processing, coverage, etc.),

Location Services, Metadata, Registries, Data Quality and other applicable standards specifically dealing with geospatial data; and

- All published data, services and applications shall include an open Application Programmers Interface (Open API). At a minimum the API shall allow searchability by the domain of geospatial search properties (e.g., contains, contained within, near, intersects etc).

Table A.1. Potential opportunities for common data, services and applications

Geospatial Data	Citation Reference	Immediate	Mid Term
Imagery	TNM, NGAC, LLR	TNM-Imagery	IFTN
Elevation/LiDAR	TNM, CSWG, NGAC, LLR	TNM-NED	NEEP
Public Streets and Roads	TNM,	TIGER	
Surface Water	TNM,	TNM-NHD	
Civil Boundaries	TNM,		
Parcels	TNM, NGAC, LLR		
Land Cover	TNM,		NLI
Geographic Names	TNM,	TNM-GNIS	
Public Land Survey System	TNM,		
Vertical and Horizontal Control	TNM,		
Federal and Native American Land Boundaries	TNM,		
Address Points	TNM, CSWG		
National Wetlands Dataset			USFWS
Geospatial Platform Services			
Cloud Computing	CSWG		
GPS receivers	CSWG		
Geocoding Service	CSWG, TNM		X
Instruction/Guidance/How-to manual		How-to manual	
Common Data Access URL			X
Remotely Sensed Land Cover Change			X
Strategy for Service Discovery			X
Geospatial Applications			
Land Cover Change	CSWG,		
Map Viewer/Visualization	CSWG, TNM	PalanterraTM	
Meta Data Editor	CSWG		
Map Viewer for Mobile Platform	TNM		X
Coastal Communities			X
Tightening Mortgage Practices			X
Cities in Transition			X

(TNM) The National Map Customer Requirements: Findings from Interviews and Surveys, OFR 2009-1222, U.S. Department of the Interior, U.S. Geological Survey, December, 2009.

(CSWG) Common Services Work Group Survey of Federal Agency Geospatial Requirements, Summary of Results, February, 2010.

(NGAC) National Geospatial Advisory Committee, Summary of key decisions/recommendations, November, 2009.

LLR) National Geospatial Advisory Committee Messaging Committee Lessons Learned and Recommendations.

Identify Priorities for New Common Data, Services and Applications

Initially, the Geospatial Platform will leverage *The National Map* requirements study and the Geo LoB surveys to provide priority recommendations for new common data, services and applications. Geospatial Platform priorities are further supported by specific studies conducted by the National Research Council (NRC), selected Federal agencies and the National States Geographic Information Council (NSGIC). The results of these studies and surveys are outlined in Table A.1.

The highest priority identified by 42% of agencies was a metadata editor that is FGDC compliant, web enabled, flexible and customizable. Also highly desired were a map tracker (33%) and an interactive mapping service (30%).

The Geo LoB surveyed agencies to identify geospatial services and applications required to meet business needs. The Geospatial Platform will leverage that survey and aim to provide the identified priority applications.

APPENDIX B. SHARED INFRASTRUCTURE, JUNE 2010

B.1. Definition

The Geospatial Platform shared infrastructure refers to joint development, operations, maintenance and/or other uses of the same IT components and investments by multiple Federal departments and agencies. Shared Infrastructure supports the hardware, networks and Platforms that common data, services and applications operate from, including portals.

A shared infrastructure for IT includes both physical and logical components that can be leveraged by multiple Federal agencies. The physical shared infrastructure includes (but is not limited to):

- Data centers and data repositories.
- Networks, including Internet and intranets.
- Host Platforms for applications, including portals and services.

Logical shared infrastructure includes (but is not limited to):

- Network routing services.
- Access control and security.

The foundation for defining a shared infrastructure is the Federal Geospatial Segment Architecture (FGSA), which is based on the Federal Segment Architecture Methodology (FSAM).

The FGSA describes the overall governing architecture for Federal geospatial investments, including business, technology, data, services and performance reference models.

The Geospatial Platform shared infrastructure identifies those components of the FGSA that can be leveraged, reused and shared by the geospatial missions of agencies and partners.

B.2. Objective/Purpose

The objective of a shared infrastructure for the Geospatial Platform is to reduce costs while improving efficiency for geospatial information sharing. A shared infrastructure allows multiple agencies and partners to leverage common investments.

In addition to reducing costs, a shared infrastructure can simplify access control requirements while improving overall security by reducing the number of physical components needed to perform operations.

Other benefits of shared infrastructure include enhancement of collaboration and services. The future state of shared infrastructure would support federated databases, which allow partners to contribute their data to larger, shared repositories where many users are able to combine and leverage them. These federated databases, for example, would create a quilt of parcel data for the Nation by combining content from State and local partners. Key drivers for this activity are identified in the Fiscal Year 2011 Budget of the United States Government, Analytical Perspectives, Section 19, Information Technology (page 321). These drivers include:

- Modernizing Federal and national IT infrastructure to be efficient and effective.
    - Centralize provisioning of IT services for non-military agencies.
    - Implement cloud computing.
    - Consolidate data centers.
- Transparency and Participation in Governance.
    - Data.gov.
    - Broadband access for Americans.
- Security and Privacy.
    - Securing government systems.
    - Identity management.
    - Protecting privacy.

In addition, the Geospatial Platform will support the President‘s Information Sharing Environment (ISE). The goal of the ISE is to assure that all authorized users can access all required data and information at any time from any authorized device.

B.3. Approach/Strategies

The Geospatial Platform will leverage common services, guidelines and strategies developed under the auspices of the Federal Chief Information Officers (CIO) Council and OMB.

These include the Federal Identify Credentialing and Access Management (FICAM) Roadmap, the Sensitive But Unclassified/Controlled But Unclassified Information (SBU/CUI) Interoperability Initiative Segment Architecture and activities to consolidate data centers across the government enterprise. Whenever possible, the Geospatial Platform will rely on shared Federal IT infrastructure to assure it is a part of the Federal mainstream IT.

Geospatial data developed by the government have broad use across a wide, interdisciplinary range of missions and needs.

To be effective the Geospatial Platform needs to recognize different needs and capabilities between customer segments, in particular, how data, services and applications are to be used. For many customers, simple view only capabilities may be required, in other cases, download of very large datasets and access to high level computing power may be required.

The Geospatial Platform shared infrastructure will address key customer segments.

B.4. Priorities

The shared infrastructure of the Geospatial Platform will develop a shared infrastructure that:

- Is robust and reliable to host and deliver common data, services and applications.
- Provides geographic information system (GIS) capability for agencies that need geospatial tools but do not have access to them.
- Provides role-based access; for example, implementing FICAM guidance for access control, including the PIVI standards for non-Federal users, is a key priority to assure data security and support the SBU/CUI information sharing environment.
- Maximizes the value of cloud computing. The Cloud Computing Model is a major component of the Federal Government's effort to achieve efficient and effective IT. The FGDC is currently working with the Federal CIO Council to develop guidance for the implementation of cloud computing.
- Supports GIS capability for offices that have integrated tools in a collaborative inter-agency mission.
- Supports the direction of the broader government IT community, including but not limited to Data.gov, USASpending.gov, Broadband for Americans, identity management, data center consolidation and National Information Exchange Model (NIEM).

The Geospatial Platform will develop an understanding of "as-is" geospatial infrastructure and user needs to help define "to-be" infrastructure.

APPENDIX C. SEGMENT ARCHITECTURE, JUNE 2010

C.1. Definition

Segment Architecture, as promoted through the Federal Segment Architecture Methodology (FSAM) is a process-driven approach to designing and deploying large government computing components in the context of a broader agency or government enterprise architecture. The Geospatial Platform requires such a methodology to design an —actionable|| solution architecture that can be readily deployed by agencies that collaborate on geospatial data, services and applications. The draft Geospatial Segment Architecture

Guidance[8] document uses the FSAM to explore the as-is and to-be Federal environment and identifies opportunities for the Geospatial Platform. The draft Geospatial Segment Architecture Guidance will serve as the key resource for guiding the Geospatial Platform segment architecture.

The Geospatial Platform will coordinate the architectural blueprint for a common geospatial data environment defined by business requirements. The architecture documents common business requirements that drive the identification of service and data resources of the Geospatial Platform to support mission needs of many agencies and stakeholders. The resulting solution architecture should be implemented by agencies using common, secure, interoperable and scalable open-standards based technologies.

The description and documentation of the Geospatial Platform's Segment Architecture is intended for three different audiences, Executives, Architects and Stakeholders.

The Roadmap provides Federal executives and decision makers with a common understanding of the Geospatial Platform architecture goals and principles and outlines initial activities to begin to transition to the target environment.

Future documents are envisioned to leverage and apply the resources of the Federal Enterprise Architecture (FEA). The Geospatial Profile and the draft Geospatial Segment Architecture Guidance document the needs of agency architects and stakeholder agencies, respectively, in response to Geospatial Platform requirements.

C.2. Objective/Purpose

The Geospatial Platform requires a segment architecture activity to identify solution engineering requirements from Geospatial Platform requirements into a consolidated, interoperable and open standards-based technical architecture, based upon the Foundational Principles listed below. The target technical architecture will be adopted and deployed by providers for the discovery and exchange of standards-based geospatial data, services and applications.

C.3. Approach/Strategies

Foundational Principles that form the basis for an efficient and cost-effective Geospatial Platform architecture include:

- Agreement on a common architectural style (i.e., a service-oriented architecture).
- Support for open standards for data content, format, protocol and service accessibility.
- Commitment to a design process that can be readily deployed by agencies as a solution architecture, while allowing partners to choose their own development methodology.
- Transparency of design, deployment and operation.

The Geospatial Platform segment architecture common engineering solution approach will:

- Facilitate the design, development and deployment of an effective and efficient Geospatial Platform segment architecture.
- Document common business process, service component, technology (standards), and data and performance requirements.
- Enable the execution of a practical and operational community geospatial segment architecture based on common principles as solution architectures within broader agency architectures.

The target Geospatial Platform architecture will be established as a services-oriented architecture (SOA) based on fundamentals of business focus and alignment using common, secure, interoperable and scalable open-standards based technologies. It will be designed to provide common geospatial capabilities for data, services, applications and infrastructure in support of stakeholder missions, business needs and citizen access. Using the Geospatial Segment Architecture Guidance document, the Geospatial Platform architecture will address each of the five reference models of the Federal Enterprise Architecture – Business, Data, Technical, Service Component and Performance – through application of the FSAM. When applied, these common engineering solutions provide efficiencies and economies within the respective agencies as well as contribute to the federated Geospatial Platform.

Segment Architecture activities shall support relevant current Federal priorities including accountability and transparency, Data.gov and the ISE, in the context of FEA guidance and methodologies.

Figure C.1 provides a logical view of the anticipated Geospatial Platform architecture development process. Geospatial Platform requirements are processed and integrated following the methodology presented in the Geospatial Segment Architecture Guidance. This process will help develop documented consensus on the common business processes, data and service requirements relevant to supporting the full Geospatial Platform.

An architectural blueprint can then be developed that formalizes the agreements on standardized data, services, applications and process definition that can be translated into tangible service offerings (solution architectures). The proposed architecture exists within a broader multi-tier context. The Platform focus is on the Platform tier, which is underpinned by standardized infrastructure tier capabilities (network, storage, security, provisioning) and is accessed by clients in the application tier. It is anticipated that the Geospatial Platform may define suites of standardized operations (shown by the purple horizontal line) that bundle specific requirements for infrastructure, data, process and services into service patterns that support shared business processes and can be deployed or procured in multiple locations, as needed. Examples of this may include geocoding services, catalog/registry services, visualization and data access services.

One of the core principles of a coordinated geospatial architecture, formalized in the FEA and FSAM, is the identification and adoption of relevant technical geospatial standards within the community. Specifying common standards for adoption within the Platform context include these benefits:

- Identified standards are developed by voluntary consensus standards organizations with a commitment to maintenance and review;
- Standards organization membership and requirements reflect the interests of the government in geospatial interoperability;
- Adoption of standards by industry and government increases the availability of deployed, compatible, software solutions; and
- Adoption of common standards facilitates immediate geospatial data exchange between agencies and between the government and its stakeholders.

Standards are defined in the FEA Geospatial Profile and Architecture Guidance Document annexes to specify documentation of data content and structure, metadata, format, and web service protocols for maximum interoperability within the Platform environment.

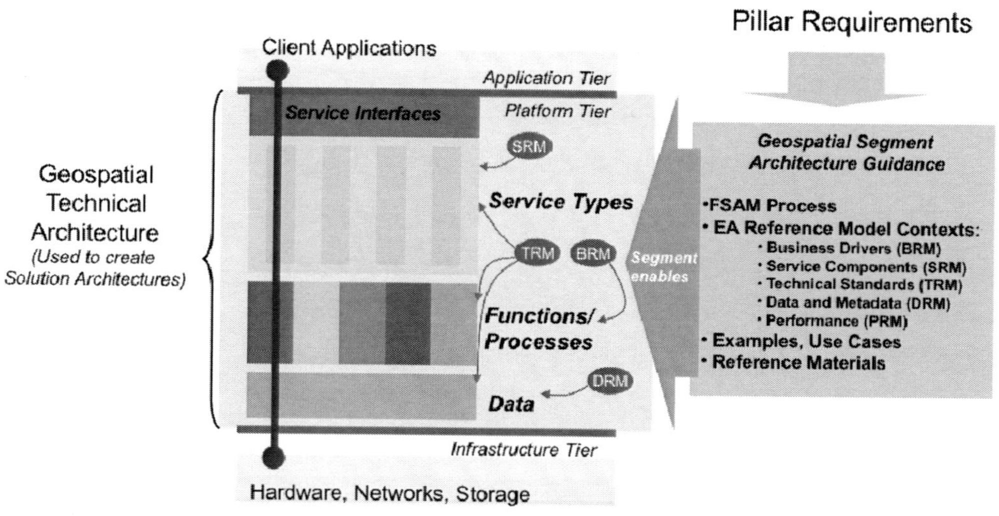

Figure C.1. Geospatial Platform conceptual architecture and role of FEA.

C.4. Priorities

The Geospatial Segment Architecture Guidance provides a context, or rubric, through which all Geospatial Platform requirements can be coordinated, resulting in a notional architecture that can be readily deployed by agencies. Potential activities/opportunities that are central to the shared architecture include:

- Formal endorsement and approval of the Geospatial Segment Architecture Guidance.
- Requirements-based identification of common, high priority geospatial business activities and processes (i.e., data, services and applications) for inclusion and/or development of the Geospatial Platform Initial Operating Capability.
- Review and update the as-is technical architecture defined by the Geospatial Segment Architecture Guidance and develop a common to-be federated infrastructure based on a SOA approach.

- Review and update the necessary geospatial open standards from the Geospatial Segment Architecture Guidance reference annexes that are required to deploy the Geospatial Platform.
- Assure that the technical architecture design supports deployment of geospatial data, services and applications using a cloud computing model as a major part of the strategy to achieve efficient and effective geospatial IT.
- Develop consensus and publish formal data content models for national geospatial data assets, including preparation of a data dictionary, data types and standard metadata to promote data acquisition, publication, conversion and use.
- Determine how the quality of data, service, or application, as defined by the customer, can be represented and used within the technical architecture.
- Document requirements for Data.gov to support registration, discovery, evaluation and visualization of geospatial data in the context of operational needs of the Geospatial Platform.
- Address requirements of Geospatial Platform data, services and applications with respect to Federated Identity, Credential and Access Management (FICAM).
- Accommodate architectural support for common applications compatible with the multi-agency Common Code Base Initiative - i.e., migrate Palanterra-based geospatial client/service functionality to a services-based approach to foster efficiency and reuse.
- Drive and anticipate geospatial capabilities in Data.gov to include geospatial and temporal search, index and cataloguing capabilities currently in Geodata.gov but required for the common USASearch.gov environment.
- Drive and anticipate augmentation of Data.gov capabilities to support standards-based service and applications documentation and discovery coupled with registered and approved geospatial data resources.
- Support development of an IT Dashboard (e.g., http://it. usaspending. gov) for alignment and reporting of cost/schedule/performance as part of the Geospatial Platform program management oversight and transparency to the geospatial community.
- Support quantification of stakeholder business demand for Geospatial Platform data, services and applications through monitoring the use of Geospatial Platform-affiliated data, services and applications.

Appendix D. Governance, June 2010

D.1. Definition

The governance structure for the Geospatial Platform is the process by which Geospatial Platform stakeholders, including customers, are afforded an opportunity to shape its structure, functions and capabilities. This governance structure includes people, policies and operational processes.

D.2. Objective/Purpose

The goal of creating a defined governance structure is to provide a standardized means to increase value to the national geospatial community, the programs across all levels of government and ultimately the American citizens who are customers.

The governance structure will involve a phased implementation approach, which will parallel phased implementation of the Geospatial Platform itself, designed to eventually engage all stakeholders and fulfill the overarching vision of the Platform.

Over time, the number and variety of stakeholders involved in governance will grow as new phases are pursued. Each phase will address a specific capability with a unique set of challenges and may have Business Models that vary across phases.

Over time, the phases will lead to a modular approach to fully building out the overall vision for the Geospatial Platform, which will have the broadest practical governance structure.

There are two primary objectives in creating the Platform governance structure. The first objective is to outline the requirements to successfully implement the governance of the Geospatial Platform in a coherent and sustainable way.

After the requirements for successful governance are identified, the primary objective will shift to operationalizing the approach by engaging Federal, State, regional, local and Tribal government stakeholders. To satisfy these objectives, a clearly defined governance approach and roles for implementation must be developed so that stakeholder groups understand the governance structure and their roles and responsibilities within the structure at each phase of its development.

The governance structure includes drivers that are crucial to developing the Geospatial Platform:

- Improving the framework to address the requirements of Federal, State, regional, local and Tribal agencies.
- Prioritizing investments based on the most widely shared and most significant business needs of government.
- Enhancing interagency cooperation.
- Enhancing intergovernmental cooperation.
- Saving taxpayers money through more effective leveraging of intergovernmental activity.
- Increasing transparency and accountability.
- Creating an integrated place-based framework for addressing public policy challenges.
- Acknowledging and aligning varying roles, capabilities and interests of parties with a stake in the Geospatial Platform.
- Building and maintaining credibility, trust and relevance amongst non-Federal stakeholders.

D.3. Approach/Strategies

Scope of Governance: Platform Implementation and Sustainment

The first objective of the governance structure will be to outline the necessary governance requirements to successfully implement the Geospatial Platform. To that end, it will be important to identify and define components where governance applies. The governance structure will ensure that aspects of Geospatial Platform work requiring governance activities are clearly defined and that the proper governance structures are in place to ensure success.

One aspect of effective governance will be to define the —rules of engagement" for the Geospatial Platform. As all participants and stakeholders may be both partners and customers, ensuring compliance is critical to Platform success.

Topics requiring compliance assurance may include:

- Data Content Standards.
- Metadata Standard.
- OMB Circular A-16 and A-16 Supplemental Guidance (Federal agencies only).
- Expectations with respect to data sharing/reciprocity.
- Expectations with respect to donating expertise within the community to achieve common goals.
- Standards for cyber security/access protocols/credentialing of Platform users to protect legitimate users.
- Managing relations with other NSDIs.
- Division of labor with respect to managing operations and maintenance of the Geospatial Platform.

While disclosure of non-compliance may often be sufficient to induce the desired behavior, the governance structure will need to define the consequences of non-compliance and the means for ensuring commitments. As such, it will be necessary to develop mechanisms to ensure that partners adhere to the agreed upon standards and —rules of engagement" included in negotiated contractual, license or service-level agreements.

To maximize effectiveness, the Geospatial Platform will leverage current interagency and intergovernmental initiatives such as Data.gov, the Geodata.gov portal and Geo LoB products and processes. Identifying and evaluating best practices and lessons learned from these initiatives will help to inform and strengthen governance of the Geospatial Platform.

After the initial governance requirements for implementing the Geospatial Platform are outlined, the primary focus will be to create a governance framework to effectively define and engage all members of the stakeholder community. This framework will help to foster more effective intergovernmental geospatial priority setting. The communities of interest to consider include:

- Federal agencies.
- State agencies.
- Regional government organizations.
- Local agencies.

- Tribal governments.
- Academia/Non-profit organizations.
- Private sector organizations.

As with other components of the governance structure, this process will unfold through a phased approach. The ultimate vision is to be a national capability with shared governance. Suggested ways to engage non-Federal entities include:

- Establish an interim Intergovernmental Subcommittee with diverse participation/representation until a full intergovernmental governance approach is developed.
- For each phase, identify means to engage specific non-Federal communities of interest with a clear stake in the phase in question. For instance, the first phase is likely to involve the current Gulf of Mexico oil spill, so the Geospatial Platform could work to engage coastal states in the Gulf of Mexico, academics, etc.
- Create appropriate means to gather customer requirements for each phase that will include targeted non-Federal participation. This could potentially be coordinated through the National Geospatial Advisory Committee (NGAC).

Outcomes include:

- Better public policy decisions.
- Faster public policy decisions.
- Better integrated public policy decisions.
- More efficient use of government funds.

Implementation of Governance: Implementation Strategy

A phased governance approach will be developed and implemented to build toward the Geospatial Platform's overarching vision. Each phase will contribute to reaching the overarching vision like independently created modules that nonetheless fit together because each was conceived with the broader vision in mind. Each phase may address a different problem with a unique set of challenges so that different Business Models could be developed and deployed to address the problem. Eventually, enough different problems will be identified and means to address those problems will be developed to lead toward a unified intergovernmental governance solution to manage the realized overarching vision. Generally, each phase should be progressively more ambitious so that lessons learned from previous phases can inform the risk management approaches in the next phase. The end goal is to incrementally develop a Platform capability that adds value in a wide variety of circumstances. Generally, each phase would:

- Define an offering that produces or delivers something to solve a specified business problem.
- Define the business institution that actually produces and manages the offering.
- Define the customers of the offering.

- Provide for internal management of the business institution to comply with applicable Federal IT, acquisition, performance, human resources and financial policies and reporting.
- Comply with the intent of the President's 2011 Budget language.

Outcomes include:

- Overarching vision for the creation of the Geospatial Platform.
- Phased approach to reach the vision for the Geospatial Platform.

Governance Approach: Governance Strategy Principles

Because the Geospatial Platform describes new processes for interactions within the community, a clearly defined a strategy for these interactions is an important component of success. This section discusses concepts that are important to follow in creating the Geospatial Platform governance structure, including specific roles and responsibilities.

First, the Geospatial Platform will operate under the federated approach. While this approach necessarily respects agency-specific statutory authorities, it nonetheless sets expectations that agencies will exercise their discretion within statutory parameters consistent with the rules of engagement for the Platform, so as to increase the overall efficiency and effectiveness of the Federal Government. It also allows for joint funding to meet operational needs. An example of this approach is the Alaska Elevation data effort, which is a joint effort of the Department of the Interior (U.S. Geological Survey and Bureau of Land Management), Department of Defense (U.S. Northern Command and National Geospatial-Intelligence Agency), U.S. Department of Agriculture, National Oceanic and Atmospheric Administration and the State of Alaska.

Second, the Geospatial Platform, initially, will capitalize on existing geospatial institutions and clearly define the future roles of existing governance bodies. The primary focus areas of Geospatial Platform governance will be to address geospatial policy, operation and coordination with Federal and non-Federal stakeholders. To those ends, the FGDC Steering Committee will serve as the primary policy body. The FGDC Executive Committee will serve as the senior management operational and coordination authority for that policy body. The Intergovernmental Body (Section 5) will serve as the primary external advisory body to establish a mechanism for non-Federal intergovernmental coordination.

Third, the governance strategy must be —workable|| and allow for performance-based evaluation (Section 5.6) of geospatial production against formal requirements. The Geospatial Platform will identify the criteria for performance-based evaluation. Having performance-based evaluation criteria compared to an established baseline allows for more informed identification of needs and performance for all managers within the community.

Outcomes include:

- Clear understanding of how each agency relates to the governance structure of the Geospatial Platform.
- Explicit expectations of each agency with respect to compliance with Platform rules of engagement.

- Federal, State, regional, local, and Tribal governments are identified and satisfied to the extent possible.
- Degree of compliance with intent of the President's 2011 Budget language to address the requirements of State, local, and Tribal governments.

Governance Roles: Entities

A key component of the Geospatial Platform governance structure will be clearly defined roles and responsibilities for stakeholder groups. The Geospatial Platform represents a new business institution, which will require the dedication of full-time support staff, as well as other forms of contribution. This section identifies geospatial communities that will facilitate the updated Geospatial Platform governance structure approaches and strategies.

FGDC Steering Committee/Executive Committee

The FGDC Steering Committee will serve as the primary policy body. The FGDC Executive Committee will serve as the senior management operational and coordination authority for that policy body.

FGDC Coordination Group

The FGDC Coordination Group will help shape the requirements for the Geospatial Platform based on agency-specific needs and priorities and will assess technical feasibility of proposed offerings. The FGDC Coordination Group will be the primary operational representatives to ensure their agencies are fulfilling their roles.

Geospatial Platform Managing Partner

Because the Geospatial Platform represents a new business institution, new capabilities will be required to support the new institution. A new body, the Geospatial Platform Managing Partner, will implement direction, recommend and/or assign resources and create and track performance metrics for upward reporting for the Platform. The focus of the Geospatial Platform Managing Partner needs to be aligned with the mission and vision of the Geospatial Platform as described in the Roadmap and be answerable to the FGDC Steering Committee, its Executive Committee and ultimately OMB.

Intergovernmental Subcommittee

The to-be state of the Geospatial Platform is a national effort with participation from State, regional, local, Tribal and non-government stakeholders. Therefore, the Geospatial Platform will quickly establish an intergovernmental subcommittee with representation from all stakeholder groups.

This is an interim solution, and will be replaced by an appropriate end-state entity as a framework for regular and ongoing coordination with these external partners and stakeholders is developed.

Outcomes include:

- Clear policy or charters for each entity involved in governance of the Platform.
- No ambiguity or misunderstanding of the roles and responsibilities for each stakeholder community.

D.4. Priorities

The first priority of the governance structure will be to establish the requirements to implement the Geospatial Platform.

After the requirements for successful governance are outlined, the primary objective will shift to working to fully engage Federal, State, regional, local and Tribal agencies and with private and non-profit stakeholders as appropriate. The priorities for the governance structure are:

- Platform implementation and sustainment

The initial priority will be to establish the requirements necessary to implement the Geospatial Platform. In order to do this, the overall vision for the Platform needs to be developed. Once the overall vision for the Platform is developed, the phases that will build into that vision need to be developed. This will also entail defining in advance how the different phases will eventually integrate with one another before they are built.

After the development of the initial vision for the Geospatial Platform and implementation plan, the focus will shift to other necessary governance infrastructure for implementation such as identifying —touch-points|| with other Platform Pillars, defining the rules of engagement, identifying means for leveraging intergovernmental initiatives and defining a funding model for the Platform.

- Federal and non-Federal involvement in the Platform

Once relevant parties have defined initial requirements to implement the Geospatial Platform, the second priority will be to organize Federal partners (through the FGDC Steering Committee and Executive Committee) and non-Federal partners (which the FGDC Steering Committee would facilitate through NGAC, National States Geographic Information Council (NSGIC), National Association of Counties (NACO), etc.).

After organizing stakeholders, exploring options for an intergovernmental needs assessment will be the next priority. To carry out the needs assessment, the governance structure must establish common terms and definitions and enforce consistent adherence to those terms. Only with such intergovernmental consistency will a survey of the intergovernmental geospatial community be useful. A qualitative assessment in FY 2011 using focus groups or doing an analysis of the data available through the 50 States Initiative will provide some initial information. After the definitions of a more robust set of terms are established, a quantitative survey would be conducted in FY 2012 and regularly maintained thereafter. In order to define the Federal agency needs, it may be necessary to modify OMB Circular A-11 so that it is for the first time possible to track Federal geospatial financial obligations through the object class structure in Federal agency financial systems.

Outcomes include:

- Organization of Federal partners.
- Engagement of non-Federal partners.
- Integration of Geospatial Platform priorities into FY 2012 budget cycle.

- Improved understanding of intergovernmental needs.

APPENDIX E. PORTFOLIO MANAGEMENT, JUNE 2010

E.1. Definition

Definition

Geospatial Platform portfolio management is defined as the set of processes used to prioritize and select assets, and allocate resources of the Geospatial Platform.

This process also supports management and performance evaluation of geospatial assets to which resources are applied to help ensure the appropriate combination of assets are available to best accomplish organizational goals and objectives of customers. The Geospatial Platform portfolio management process supports investment decisions for four categories of geospatial assets: data, services, applications and infrastructure.

Scope

The Geospatial Platform portfolio consists of geospatial assets that are deemed —significant/critical/necessary for meeting crosscutting (i.e., common mission needs) of the Federal Government or its State, Tribal and local partners and stakeholders.

Assets are selected (or rejected) for inclusion in the Geospatial Platform portfolio through a portfolio management process that includes the application of selection criteria to assess how the assets meet common requirements and business needs across customers, and how well the assets support the strategic goals and objectives of the Geospatial Platform.

E.2. Objective/Purpose

The overarching objective for the Geospatial Platform portfolio management process is to maximize enterprise value by investing in the right assets at the right time and thus obtain the best possible strategic impact of each investment.

Geospatial Platform portfolio management will support senior level decision making by:

- Guiding the priority setting, selection, and balancing of portfolio assets to ensure that the overall investments align with the strategic goals and highest organizational priorities of the enterprise and partners.
- Allowing timely assessments of asset selection, prioritization, and performance, as well as the identification of any portfolio-level issues and/or risks that might impact performance.
- Enabling transparent, timely and consistent communications to decision makers and the public on the progress for completing the Platform, any changes to the Platform, and impacts of these on the Geospatial Platform portfolio.

Specific objectives for the portfolio management process are:

- *Objective 1:* Adopt a portfolio management framework that enables the Managing Partner to select assets that will best support projected Geospatial Platform needs, capabilities, and projected outcomes and disinvest in assets that do not support the desired outcomes for the Geospatial Platform.
- *Objective 2:* Identify the most important crosscutting strategic endeavors and outcomes for the Geospatial Platform.
- *Objective 3:* Identify and adopt a series of best practices and tools for each step of the portfolio management process.
- *Objective 4*: Discover and adopt appropriate measures to control, monitor, and evaluate performance of existing and/or planned assets within the Geospatial Platform portfolio.
- *Objective 5:* Determine the most appropriate maturity, investment and readiness models to use as part of the Geospatial Platform portfolio management process.
- *Objective 6:* Acquire/adapt a portfolio management system for enterprise level management and as a common service available to individual assets for tracking.
- *Objective 7:* Apply a portfolio management process framework so an initial Geospatial Platform portfolio is chosen and subsequently controlled and monitored for performance.
- *Objective 8:* Evolve the content of the Geospatial Platform portfolio over time as available asset maturity increases and adjusts to strategic goals and events necessitating change.

Portfolio Management Drivers

Drivers for developing the Geospatial Platform portfolio management process include:

- Improving management of Geospatial Platform capabilities and assets to meet the business needs of the customers, including using resources more efficiently in an era of decreasing budgets across all levels of government.
- Providing access to critical geospatial data, services and applications currently inaccessible or difficult to obtain.
- Responding more promptly and accurately to inquiries made to Federal agencies on the scope, cost and completion progress for completing the NSDI.
- Incorporating portfolio and performance management approaches to leverage resources and better support outcomes for mission-based goals, transparency and accountability.
- Ensuring compliance with the transparency, open government and effective and efficient information sharing initiatives.

E.3. Approach/Strategies

Establishing a portfolio management process gives the Geospatial Platform a clear structure for transforming current business priorities into a portfolio of assets needed to support mission execution. The Geospatial Platform portfolio management process will adhere to best practices and will utilize management tools that can be adapted to reflect the organizational governance of the Geospatial Platform. In addition, the portfolio management process will include a geospatial investment review process through which the FGDC Steering Committee considers existing and planned assets while assessing future initiatives and changing business drivers.

All portfolio asset investments will follow a lifecycle management approach (for data, view the A-16 Supplemental Guidance; for services, applications and infrastructure, the OMB Circulars A-11 and A-130). Geospatial asset selection criteria are critical for effectively and efficiently identifying the components of the Geospatial Platform portfolio and should be reflected in a lifecycle management framework from the inception of the Geospatial Platform.

In addition, best practices and associated decision tools will be established and adopted for each step of the portfolio management process and the lifecycle management approach. Geospatial Platform asset capability, maturity, readiness and performance measures will also be developed, reflecting the desired outcomes of OMB, Platform partners and customers and the FGDC Steering Committee.

To the greatest extent possible, Geospatial Platform portfolio management will utilize existing portfolio and lifecycle management frameworks, maturity models, best practices and tools, as well as dashboard and balanced scorecard approaches. Used in tandem with standard management approaches, these tools and analysis processes can help evaluate and manage the health of the Geospatial Platform portfolio assets. Examples of existing standards for project, program and portfolio management as well as best practices, tools, standard operating procedures and IT investment dashboards that may be adopted and/or adapted for use in the portfolio management include:

- Requirements within Exhibit 300/Capital Planning and Investment Control (CPIC) for Select, Monitor and Evaluate.
- OMB Circulars A-16, A-11, A-130.
- Processes outlined within the A-16 Supplemental Guidance.
- The Performance Institute Standard for Portfolio Management (ANSI/PMI 08-003-2008).
- Information Technology Infrastructure Library (ITIL).
- The U.S. Government Accountability Office (GAO) IT investment management maturity model.

E.4. Priorities

The portfolio management priorities include:

- Formal endorsement and approval of the A-16 Supplemental Guidance.
- Initial inventory and selection of geospatial data, services and applications (See Common Data, Services and Applications priorities).
- Acceptance, adoption and implementation of A-16 Supplemental Guidance lifecycle management process and reporting for National Geospatial Data Assets (NGDA).

- Complete proposals for services, applications and infrastructure portfolio management processes and gain FGDC Steering Committee concurrence (see Common Data, Services and Applications priorities).
- Improve the alignment of the FGDC cross-agency investments to Administration priorities and Federal agency high value mission objectives.
- Research and consider the American National Standards Institute (ANSI) Standard for Portfolio Management as the Geospatial Platform framework for portfolio management governance.
- Support Administration priorities for openness, transparency and participatory government.
- Complete readiness assessment of various components so that the Geospatial Platform optimizes investments.

APPENDIX F. FEDERAL GEOGRAPHIC DATA COMMITTEE STRUCTURE AND ROLES

F.1. Federal Geographic Data Committee (FGDC) and Work Group Structure

The Federal Geographic Data Committee (FGDC) is an organized structure of Federal geospatial professionals and constituents that provide executive, managerial, and advisory direction and oversight for geospatial decisions and initiatives across the Federal Government. In accordance with OMB Circular A-16, the FGDC is chaired by the Secretary of the Department of the Interior with the Deputy Director for Management, OMB as Vice-Chair. The FGDC's structure is represented in Figure F.1.

The acronym *FGDC* is often used interchangeably to refer to the Steering Committee; the Office of the Secretariat (OS); or the committee structure as a whole, causing confusion. The Federal Geographic Data Committee is the Steering Committee, which is made up of senior Federal agency officials, and directs, approves, and oversees the activities of all the supporting groups, subcommittees, and work groups. The FGDC OS is an office hosted in the U.S. Geological Survey that provides staff support for myriad FGDC activities and initiatives, and fulfills a number of key supporting roles.

Steering Committee

The FGDC is governed by a *Steering Committee,* which is the policy-level interagency group whose central focus is to provide executive leadership for the coordination of Federal geospatial activities between, among, and within agencies by establishing policy and providing guidance and direction to the member agencies. The Steering Committee is responsible for overseeing OMB Circular A-16 related activities and the implementation of the National Spatial Data Infrastructure (NSDI). The FGDC Chair and Vice-Chair lead this committee, which is made up of Senior Agency Officials for Geospatial Information (SAOGIs) and has representatives from Federal organizations, including the Executive Office of the President and Cabinet-level and independent Federal agencies.

Executive Committee

The FGDC *Executive Committee* is a subset of the Steering Committee and provides advice and guidance to the FGDC Chair and the Vice Chair on major Federal geospatial priorities and initiatives.

The FGDC Chair and Vice-Chair lead this committee, which has representation from OMB and the seven Federal agencies with the largest investments in geospatial technologies.

Coordination Group

The FGDC *Coordination Group* provides advice on the day-to-day business of the FGDC to facilitate interagency coordination and implementation of the NSDI at the operational level. The Coordination Group oversees and provides the functional leadership for the FGDC subcommittees and working groups, as well as the Geospatial Line of Business (Geo LoB) Work Groups.

The Coordination Group is Co-Chaired by the Executive Director of the FGDC Office of the Secretariat (FGDC OS), and an elected Federal member of the Coordination Group. Non-Federal collaborating partners participate in most of the Coordination Group meetings and work on subcommittees and work(ing) groups.

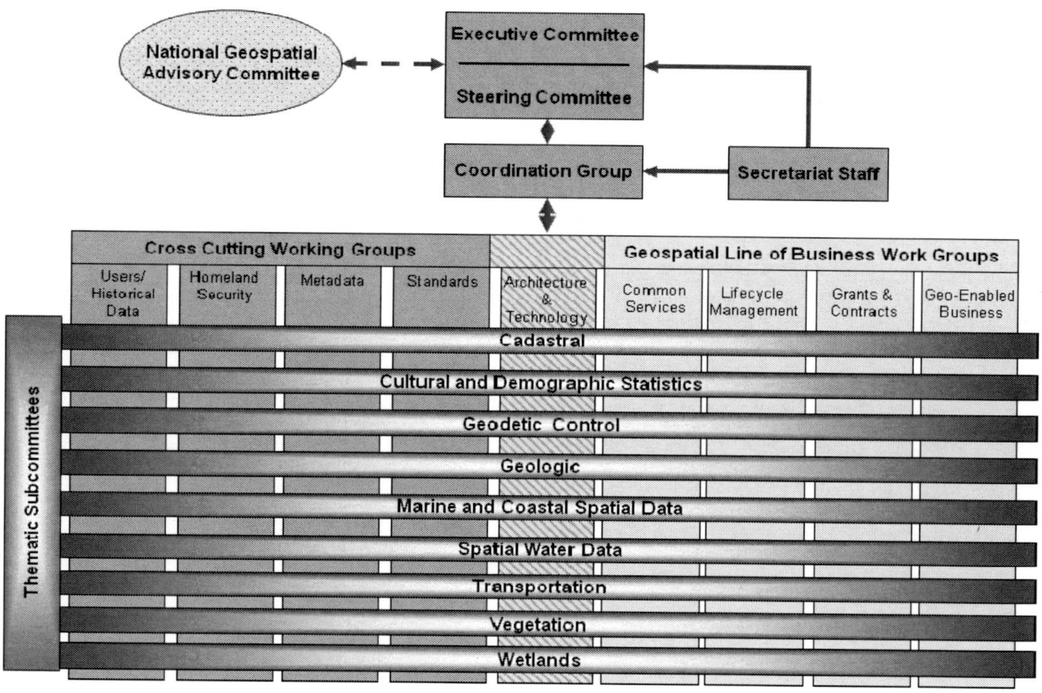

Figure F.1. FGDC Structure.

FGDC Subcommittees, Working Groups, and Geospatial Line of Business Work Groups

The FGDC structure includes agency-led working groups and subcommittees and Geo LoB Work Groups.

FGDC Thematic Subcommittees - OMB Circular A–16 (2002) identifies 34 data themes of national significance and assigns responsibility for each of the data themes to one or more Federal agencies. FGDC thematic subcommittees are established for nine of the data themes to address data theme issues. Federal agencies have responsibility for, and lead, the thematic subcommittees.

FGDC Working Groups - FGDC working groups crosscut the subcommittees and focus on issues common to, or supporting the NSDI data themes, such as standards and common services.

Geospatial Line of Business Work Groups - The Geo LoB Work Groups were developed to support this electronic government (E-Gov) initiative. The FGDC has been assigned responsibility for these work groups by the OMB.

The Geo LoB Work Groups focus on cross-agency geospatial issues related to geospatial policy, business requirements, and business management processes.

National Geospatial Advisory Committee

The National Geospatial Advisory Committee (NGAC) was established under the Federal Advisory Committee Act (FACA) and is sponsored by the U.S. Department of the Interior. The NGAC is an advisory body that provides advice and recommendations on Federal geospatial policy and management issues and is a forum to convey views representative of partners in the geospatial community.

NGAC membership includes representatives from 28 Government and nongovernmental organizations. The committee holds public forums to discuss geospatial activities and solicits input from State, Tribal, regional, and local governments, academic institutions, and the private sector.

FGDC Office of the Secretariat (FGDC OS)

The FGDC OS is an office located within the U.S. Geological Survey in the Department of the Interior. The FGDC OS staff provides support for all components of the FGDC structure.

The FGDC OS is not a policy making body but rather supports the efforts of the Steering Committee, Executive Committee, and Coordination Group. The FGDC OS staff has numerous roles and performs numerous tasks in support of the FGDC. The FGDC OS:

- Acts as Managing Partner for the Geo LoB, E-government initiative, on behalf of DOI.
- Is the Designated Federal Official for the NGAC FACA committee.
- Provides staff support for the FGDC Chair, Vice-Chair, and their designees.
- Initiates, participates, and or leads FGDC committees, subcommittees and working groups.
- Drafts policies and procedures for consideration and approval by the Coordination Group, the Executive Committee, and the Steering Committee.
- Provides support to the NGAC and its subcommittees.
- Provides project management support for FGDC initiatives.
- Administers the FGDC standards program.
- Administers the NSDI Cooperative Agreements Program (CAP).

- Administers the FGDC International Spatial Data Infrastructure program.
- Manages the NSDI training and outreach program.
- Maintains the FGDC website and membership lists.
- Manages all administrative requirements associated with scheduling and conducting meetings.
- Undertakes staff analysis, technical development, and other activities on behalf of the Coordination Group.

Collaborating Partners

The FGDC solicits the involvement of public interest groups who participate within the committee structure to ensure that their needs are included in the developing NSDI.

These collaborating partners include State, Tribal, and local governments; academic institutions; and a broad array of private sector geographic, statistical, demographic, and other business information providers and users. The NSDI strives to build upon local data wherever possible.

Collaborating partnerships are open to public, private, and nonprofit organizations whose missions are complementary to the mission of the FGDC.

End Notes

[1] OMB Memo 09-28 is available at http://www.whitehouse.gov/omb/assets/ memorandafy 2009 /m09-28.pdf

[2] President's Budget, Fiscal Year 2011, "Analytical Perspectives, Special Topics, Information Technology" (p. 325)

[3] President's Budget, Fiscal Year 2011, "Analytical Perspectives, Special Topics, Information Technology|| (p. 325) (www.whitehouse.gov/omb/ budget/fy2011/assets/topics.pdf)

[4] For more information about Federal Enterprise Architecture visit www. whitehouse.gov/omb/e-gov/fea/

[5] To geo-enable is to apply geospatial capabilities to a business process in order to establish the authoritative spatial location of business data, and enable contextual spatial analysis.

[6] http://www.whitehouse.gov/omb/budget/fy2011/assets/topics.pdf

[7] Detailed information on data portfolio management can be found in the OMB Circular A-16 Supplemental Guidance www.fgdc.gov

[8] The draft Geospatial Segment Architecture Guidance is available at: http://www.fgdc. gov/ geospatial-lob.

In: Federal Geospatial Information
Editors: Andrew Halpern and Mark Troupp
ISBN 978-1-62257-572-5

Chapter 4

FEDERAL GEOGRAPHIC DATA COMMITTEE: 2011 ANNUAL REPORT*

Federal Geographic Data Committee

MESSAGE FROM THE FGDC CHAIR

I am pleased to present the Federal Geographic Data Committee's (FGDC's) annual report for 2011, describing the activities and accomplishments of the FGDC partner agencies over the past year. 2011 has been a year of accomplishment and progress, with significant advances in the development of collaborative geospatial data and technology. We look forward to continuing this progress in the coming year.

Over the past year, the partner agencies of the FGDC have made significant progress on a number of fronts. For example, the FGDC has made significant progress in conceptualizing and beginning implementation of the Geospatial Platform initiative. The Geospatial Platform, which was described in the President's fiscal year 2011 budget, will be an accessible, reliable, and cost-effective collection of data, applications, and services that will provide a foundation to help address the White House's priorities for data sharing, transparency, and collaboration. In 2011, the FGDC developed a conceptual roadmap for the Platform and received very helpful feedback from external partners on the National Geospatial Advisory Committee and through a public outreach process. The FGDC also began the implementation of the first operational phase of the Platform in 2011. We are excited by the prospects for this new initiative, and we look forward to working with our partners in the coming year to realize the vision for the Platform.

The FGDC has also provided support for key White House initiatives. For example, the FGDC worked with the Office of Management and Budget (OMB) to develop new Supplemental Guidance to OMB Circular A–16, "Coordination of Geographic Information and Related Spatial Data Activities." The new guidance outlines a process to use portfolio

* This is an edited, reformatted and augmented version of a Federal Geographic Data Committee publication, dated 2011.

management approaches to more effectively manage Federal geospatial assets and investments. OMB issued the new guidance to agencies in November 2010.

In addition, the FGDC has made significant progress in the area of geospatial standards. The FGDC adopted 64 external standards, including many from the International Organization for Standardization (ISO) and the Open Geospatial Consortium, and also adopted the U.S. Thoroughfare, Landmark, and Postal Address Data Standard. In addition, the FGDC is coordinating standardization activities with the Department of Defense. Standards are a key component of the National Spatial Data Infrastructure and we are pleased with the progress in this vital area.

Fiscal year 2012 will focus on continued collaboration to better enable public-private partnerships and investments, and continued progress on development of the Geospatial Platform and other tools and data for the benefit of the agencies and public we serve. We are also rejuvenating the FGDC Steering Committee and working with senior officials in our partner agencies to advance national geospatial programs. We look forward to a productive year.

I thank all of you who have contributed to the success of the FGDC over the past year.

Sincerely yours,
Anne Castle FGDC Chair
Assistant Secretary of the Interior for Water and Science

FGDC Accomplishments for Fiscal Year 2011

1. Geospatial Line of Business

The Geospatial Line of Business obtained the Office of Management and Budget (OMB) endorsement for the OMB Circular A–16 Supplemental Guidance, which includes processes, roles, and responsibilities for managing National Geospatial Data Assets datasets and themes, and a repeatable process for adjusting OMB Circular A–16 Appendix E that describes the National Spatial Data Infrastructure geospatial data themes.

2. Portfolio Management

The Geospatial Line of Business continues to make strides towards portfolio management through the development of an implementation strategy and agency implementation plan templates. Another significant accomplishment is the development of a current inventory of National Geospatial Data Assets.

3. Geospatial Platform

The "Modernization Roadmap for the Geospatial Platform" document, completed in March 2011, provides a framework to describe how the concepts and goals of the Geospatial Platform were developed and released. This "Roadmap" concludes the upfront planning process for the Geospatial Platform and paves the way for implementation activities.

The redesigned Geospatial Platform website (*www.geoplatform.gov*) was released. The Platform is a managed portfolio of common geospatial data, services, and applications contributed to and administered by authoritative sources and hosted on a distributed, cloud-based infrastructure. The Geospatial Platform is used by government agencies and partners to meet their mission needs and the broader needs of the Nation.

4. Geospatial One-Stop and the Clearinghouse Network

In fiscal year 2011, the Geospatial One-Stop (GOS) project continued to focus on integrating its geospatial data with Data.gov. Connectivity between GOS and Data.gov takes place using GOS Web service capabilities that enable searches for geospatial data through applications beyond the GOS portal. During fiscal year 2011, the GOS data collection grew to over 678,000 records, of which over 391,000 are shared through Data.gov. The GOS to Data.gov integration is in its final stages and many of the capabilities of GOS will be integrated into Data.gov in the coming months (early fiscal year 2012).

At that time, the GOS portal, *www.geodata.gov*, will cease to exist, and visitors to that portal will be redirected to a geospatially enabled page for Data.gov.

5. National Geospatial Advisory Committee

The National Geospatial Advisory Committee (NGAC) was established by the Department of the Interior to provide external advice and recommendations to the FGDC. During the past year, the NGAC has analyzed and provided feedback and recommendations on the Geospatial Platform Initiative, *The National Map*, geolocation privacy, the National Land Imaging Program, Transportation for the Nation, and the Federal Aviation Administration's laser pointer issue. The NGAC has also adopted white papers on local government Geographic Information System best practices and best practices for interagency data sharing.

6. Standards

The FGDC endorsed the United States Thoroughfare, Landmark, and Postal Address Data Standard. The standard covers street addresses, which are the location identifiers most commonly used by all levels of government and the public. Addresses are critical for administrative records, emergency and first response, research, marketing, mapping, geospatial information systems, routing and navigation, and law enforcement.

7. Geospatial Metadata

The FGDC led a community effort to generate and compile input to the "ISO 19115: Geographic Information – Metadata" five-year review. The information collected from the community provided critical insight into issues experienced when applying the standard to real world geospatial data and operations. A revised draft international standard is expected to be released in November 2011 as "ISO 19115-1 Geographic Information – Metadata - Fundamentals."

8. Fifty States Initiative

Nine States completed and began implementing strategic and (or) business plans that facilitate the coordination of geospatial programs, policies, technologies, and resources in their States and with the Federal Government.

9. NSDI Training Program

The FGDC partnered with all sectors of government, academia, Tribal Nations, and the private sector to provide National Spatial Data Infrastructure (NSDI) related training. Cooperative agreements, memorandums of understanding, contracts, and cost sharing were utilized to design and lead workshops, develop training materials for the classroom and the Web, and support travel and logistics. In most cases, Federal agencies, including the National Oceanic and Atmospheric Administration and the U.S. Geological Survey, provided trainers and training materials; State and local governments and U.S. Geological Survey Geospatial Liaisons provided logistical support; and private-sector companies sometimes contributed facilities and supplies. By sharing resources, these organizations were able to collectively address the NSDI training needs of their communities.

Success Story Monitoring Trends in Burn Severity Program

Challenge: There is a need to provide summary information regarding the effects of large wildfires to agency leaders, elected officials, and the general public. The Wildland Fire Leadership Council (WFLC), which implements and coordinates National Fire Plan (NFP) and Federal Wildland Fire Management Policies, adopted a strategy to monitor the effectiveness and effects of the National Fire Plan and the Healthy Forests Restoration Act (HFRA). One component of this strategy is to assess the environmental impacts of large wildland fires and identify the trends of burn severity on all lands across the United States.

Action: Monitoring Trends in Burn Severity (MTBS), a project initiated in 2005, is conducted through a partnership between the U.S. Department of Agriculture's Forest Service Remote Sensing Applications Center (RSAC) and the U.S. Geological Survey's National Earth Resources Observation and Science (EROS) Center.

The project's main objective is to provide a national analysis of trends in burn severity for the National Fire Plan.

The geospatial and tabular data generated by MTBS can be used at a range of spatial, temporal, and thematic scales and are intended to meet a variety of information needs that require consistent data about fire effects.

Result: The project generates a suite of geospatial data products for each fire including burned area perimeters, 30 meter pre/post-fire Landsat imagery subsets, burn severity data, and summaries based on Geographic Information System (GIS) analysis with key geospatial data layers. To date, over 13,000 fires have been mapped by the MTBS project. The number of fires that will be mapped during the course of the project is estimated to exceed 15,000. MTBS developed a public-facing website (*www.mtbs.gov*) with robust data delivery capabilities. Users can pull data bundles at multiple scales, including national, regional, and fire level. In addition to accessing geospatial data bundles, the website also allows users to query the extensive MTBS database. Visualization and analysis tools are provided to assess fire trends at various scales and in the context of key geographic strata. Tabular data for specified queries can also be downloaded for further, customized analysis.

THE NATIONAL BROADBAND MAP: COLLABORATION WORKS

Introduction

A unique collaboration among Federal agencies, State governments, and the private sector paved the way for the Nation's first geospatial dataset of broadband availability and the first public, searchable nationwide map of broadband availability. Launched on February 17, 2011, the National Broadband Map, available at *www.broadbandmap.gov*, is an unprecedented searchable database of information on high-speed Internet access. The Map and its accompanying dataset fill an important policy gap, providing—for the first time—information at a detailed geographic level about broadband availability across the country. The products also represent the culmination of two years of committed partnership by State agencies and non-profit organizations, public and private sector broadband providers, and the Federal Government. The effort exemplifies the goals of the Office of Management and Budget (OMB) Circular A–16, which established a coordinated approach to developing the National Spatial Data Infrastructure (NSDI), including the development of national data and collaborative partnerships.

The National Broadband Map supports efforts to expand broadband access and adoption in communities at risk of being left behind in the 21st century economy and to help businesses and consumers seeking information on their high-speed Internet options. It does this by providing better data with which consumers, business leaders, and policymakers can make more informed decisions. The first dataset resulting from this Federal-State-private sector partnership included 25 million searchable records showing where broadband Internet service is available, the technology used to provide the service, the maximum advertised speeds of the service, and the names of the providers offering the service. In addition, they include the type and speed of the service to which community anchor institutions—such as libraries, schools, and hospitals— subscribe. The interactive and searchable website displaying the data allows users to search by address to find the broadband providers and services available in the corresponding census block or road segment, view the data on multiple maps, or use other interactive tools to compare broadband across various geographies, such as States, counties, or congressional districts.

A Call to Action

The National Broadband Map implements the joint purposes of the American Recovery and Reinvestment Act (Recovery Act) and the Broadband Data Improvement Act (BDIA), which envisioned a comprehensive program, led by State entities or non-profit organizations working at their direction, to facilitate the integration of broadband and information technology into State and local economies. In calling for the National Broadband Map, the Recovery Act required the National Telecommunications and Information Administration (NTIA) to:

> Develop and maintain a comprehensive nationwide inventory map of existing broadband service capability and availability in the United States that depicts the geographic extent to which broadband service capability is deployed and available from a commercial provider or public provider through each State. Not later than 2 years after the date of the enactment of this Act, the Assistant Secretary shall make the broadband inventory map developed and maintained pursuant to this section accessible by the public on a World Wide Web site of the National Telecommunications and Information Administration in a form that is interactive and searchable.[1]

In addition, the Recovery Act provided the funding for the Map and linked the development of the Map to Public Law 110-385, the State Broadband Data and Development Grant program, a key component of BDIA.

With only 24 months to meet its congressional deadline, NTIA and its partners worked quickly. Less than five months after Congress passed the Recovery Act, in July 2009, NTIA released program rules through a Notice of Funds Availability. Making one grant award available per State, Assistant Secretary Lawrence E. Strickling noted that the data States gathered would be "used to create the national broadband map, and in addition, States will be able to create their own broadband maps tailored to their specific needs."2 NTIA awarded the grants beginning in October 2009 and all but a few of the 56 grantees had funding in-hand by the end of January 2010, just 13 months prior to the statutory deadline for publication of the National Broadband Map.

Critical Partnerships

Collaboration between grantees and broadband providers was critical to the timely launch of the National Broadband Map. In nine short months between January and October 2010, States—many with no initial broadband availability dataset from which to expand—completed two rounds of data collection, ultimately gathering information from approximately 3,400 State-based broadband providers representing more than 1,650 unique broadband providers. Each grantee first developed relationships with broadband providers across its State to explain the program's purpose. Identifying providers and communicating the need for the information was a time and resource intensive task, and grantees used multiple sources to identify the potential community of broadband providers. In many cases, providers and NTIA's grantees worked out specific agreements for the disclosure of information. This provider-offered data resulted in an important base of information and required a significant voluntary investment of resources by broadband providers across the country.

Many grantees also chose a collaborative governance model that supported the successful collection and verification of data. In many cases, the State teams included staff with Geographic Information Systems (GIS), economic development, and telecommunications backgrounds.

Broadband providers have only recently begun to organize telecommunications data in a spatial format or relative to census geography; as a result, grantees used significant GIS resources, converting data into the census block or road segment level NTIA required for the program.

Grantees also used a range of analysis and verification methods, from drive-testing wireless broadband service across highways to performing field investigations. After reviewing existing data available at regulatory commissions or meeting with State and local officials, many grantees returned to broadband providers, large and small, to confirm data or suggest more accurate depictions of their service areas.

Grantees delivered their State datasets to NTIA on October 1, 2010, only 4.5 months before the congressional deadline. At this point, another partnership took center stage: NTIA and the Federal Communications Commission (FCC). As two Federal agencies at the forefront of assessing and expanding broadband deployment in the United States, their expertise in these areas made them natural partners for the National Broadband Map. The agencies together developed requirements for the intake, review, and publication of data at the national level. These requirements resulted in the award of three contracts by the FCC, approximately concurrent with the grantees' data submissions. With State data, FCC and NTIA staff support, and contractual resources, the two agencies were able to meet the tight statutory deadline for launching the map.

While the Federal Government provided technical assistance and ideas to support grantees during the data collection and validation process, grantees served as a critical technical support mechanism for each other. Grantees have developed best practice methodologies for evaluating changes in provider participation, defining the composition of various classes of broadband technology types, and evaluating satellite coverage, just to name a few examples. National groups that include a large number of grantees also support this effort. The data model in use by grantees was first developed by members of the National States Geographic Information Council (NSGIC) and has continued to be improved through grantee collaboration. The National Association of Regulatory Utility Commissioners, the Rural Telecommunications Congress, and NSGIC have all devoted tracks at their regular member meetings to discuss and share best practices among the grantees present. The unprecedented nature of this work resulted in these national member groups serving an even more important role than they might have in a more established program.

From Collection To Publication

The Map meets the requirements of the Recovery Act's mandate, complementary White House initiatives, and the needs of potential users. The Recovery Act required that the final product be comprehensive, interactive, and searchable, and display broadband service capability and availability. Consistent with the White House Open Government Initiative, the team also incorporated transparency, participation, and collaboration as necessary elements for a successful final product. With these core requirements in place, staff then considered who would use the product and designed the requirements for the National Broadband Map platform based on the needs of those potential users.

Stakeholder feedback underpinned the development of use cases for distinct groups and important website functionality. For example, policymakers represent one stakeholder group. Consumers and small businesses represent two others. Stakeholders also expressed an expectation to see their search queries quickly and in a digestible format. After reviewing these needs, NTIA and the FCC identified three core functions: search, analyze, and map. A consumer is able to search broadband options at a neighborhood level (in the case of the Map, a neighborhood equates to a census block or road segment). A researcher or an economic developer analyzes the data and is able to determine where his community ranks in comparison to similar areas. A policymaker is able to view maps that tell the story of broadband availability both across the entire country and at a granular level.

The openness and transparency of the data and process, both to drive use and to ensure accountability, align this project with the White House Open Government Initiative. While the final product allows users to view broadband data in the context of demographic information, and across multiple political boundaries, the Federal team expected that stakeholders would have additional uses for the data. For this reason, the entire National Broadband Map is available through 37 application programming interfaces (APIs) (*www.broadbandmap.gov/developer)* and through a download (*www.broadbandmap.gov/ data-download)*. These tools allow developers to leverage the data and allow researchers to analyze the data in innovative ways. Moreover, the complete technical process that the FCC and its contractors implemented to assemble the 25 million records provided by grantees into a geospatial dataset that could meet the program's needs are fully detailed on the Map's Technical Overview page (*www.broadbandmap.gov/ about/technical-overview*).

What's Next?

In only the first few months, more than 500,000 unique visitors used the website. On average, they stayed for over 10 minutes. Federal and private sector groups are consuming APIs in exciting ways and researchers and others have downloaded the data over 1,000 times and are beginning to publish their findings. On September 21, 2011, NTIA, in collaboration with the FCC, and using data that each State, Territory, and the District of Columbia (or their designees) provided, updated the National Broadband Map.

Collaboration between and among State governments, nonprofit organizations, Federal agencies, broadband providers, and many other stakeholders helped bring the National Broadband Map to fruition. Every partner in this project was committed to meeting the deadline set by Congress. This support—and the agility demonstrated daily by so many of the key partners and stakeholders—has been critical to the project meeting its goals. States are currently funded to continue updating their datasets through the end of 2014, and NTIA and the FCC plan to continue working together to update the National Broadband Map website twice a year with new data from grantees.

Through its collaboration with State and other Federal agencies, the National Broadband Map team will continue to make the project a vibrant example of the NSDI.

State of the Standards

Survey of Standards Usage

Standards facilitate the development, sharing, and use of geospatial data. The FGDC develops geospatial data standards for implementing the National Spatial Data Infrastructure, in consultation and cooperation with State, local, and Tribal governments; the private sector and academic community; and, to the extent feasible, the international community. The FGDC develops geospatial data standards only when no equivalent voluntary consensus standards exist, in accordance with Office of Management and Budget (OMB) Circular A–119 on Federal Participation in the Development and Use of Voluntary Consensus Standards and in Conformity Assessment Activities.

In 2011, the FGDC conducted its first survey on FGDC member agency use of FGDC-endorsed standards and requirements for standards. Fifteen FGDC member agencies from 11 of the 31 FGDC Departmental level members provided complete responses to the survey. It should be noted that agencies need to use only the standards applicable to their mission areas. A select subset of the survey results are presented here.

Agency Policies on Standards

Survey results show that 11 FGDC member agencies (73 percent of respondents) have developed internal guidance on geospatial standards that are applicable for their use.

Three responding agencies (20 percent) reported that their agencies have not identified standards that are applicable for their use.

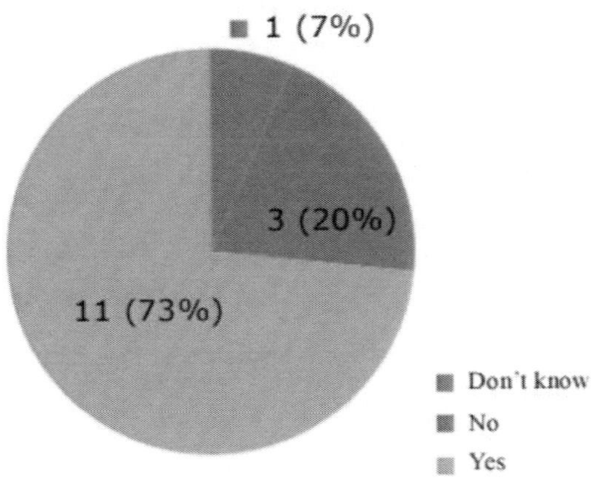

Graphic showing number of responding agencies that have identified standards for agency use.

The survey queried FGDC member agencies on whether they measured compliance with standards. Seven agencies reported that they measure compliance with standards, while an equal number reported that they do not measure compliance with standards.

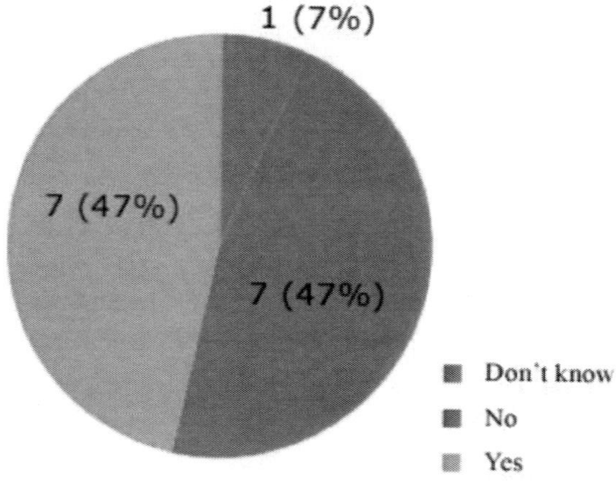

Graphic showing number of responding agencies that measure compliance with standards.

Use of Standards

The survey queried FGDC member agencies on their use of 100 FGDC-endorsed standards, including the 64 externally developed standards that the FGDC endorsed in 2010 in support of the Geospatial Platform.

Table 1 lists the ten standards with the highest level of use, based on responses. As metadata is key to geospatial data sharing through data documentation, discovery, and determining applicability of use, it is not surprising that it ranks as the standard of highest use.

Table 2 identifies the five standards with the highest number of planned implementations. Since the processes and tools to migrate to the International Organization for Standardization (ISO) metadata standard are currently a hot topic, it is not surprising to see it ranking as the number 2 "plan to use" standard.

The FGDC-endorsed standards usage survey is intended to become an annual reporting mechanism aligning with the broader OMB Circular A–16 Supplemental Guidance implementation of geospatial data portfolio management and reporting. This year's responses will serve as an initial baseline for future standards usage reporting.

FGDC Response to NIST Request for Information

The FGDC responded to a Request for Information (RFI) from the National Institute of Standards and Technology (NIST) on "Effectiveness of Federal Agency Participation in Standardization in Select Technology Sectors for the National Science and Technology Council's Sub-Committee on Standardization" that was published in the Federal Register. In this RFI, NIST invited interested parties to provide their perspectives on the effectiveness of participation by Federal agencies in the development and implementation of standards and conformity assessment activities and programs.

Table 1. The ten most widely used FGDC-endorsed standards

Standard		Number (Percentage) of responding FGDC member agencies that use the standard
1	Content Standard for Digital Geospatial Metadata (version 2.0)	12 (80.0%)
2	Geospatial Positioning Accuracy Standards, Part 3: National Standard for Spatial Data Accuracy	9 (60.0%)
3	Classification of Wetlands and Deepwater Habitats in the United States	8 (53.33%)
4	Geospatial Positioning Accuracy Standards, Part 1: Reporting Methodology	8 (53.33%)
5	INCITS 455:2009, Information Technology - Codes for the Identification of Congressional Districts and Equivalent Areas of the United States, Puerto Rico, and the Insular Areas	7 (46.67%)
6	OpenGIS (OGC) KML (version 2.2.0)	7 (46.67%)
7	National Vegetation Classification Standard (version 2)	7 (46.67%)
8	Part 6, Hydrography, FGDC-STD-014.6-2008	7 (46.67%)
9	GeoTIFF Revision 1.0	7 (46.67%)
10	Content Standard for Digital Orthoimagery	6 (40.0%)

Table 2. FGDC-endorsed standards planned for use

Standard		Number (Percentage) of responding FGDC member agencies that plan to use the standard	Number (Percentage) of responding FGDC member agencies that are using the standard
1.	United States Thoroughfare, Landmark, and Postal Address Data Standard	7 (46.67%)	4 (26.67%)
2.	INCITS/ISO 19115:2003 [R2008] ISO 19115:2003w/ISO 19115 Cor. 1:2006, Geographic information - Metadata + Corrigendum 1	6 (40.0%)	0 (0.0%)
3.	OpenGIS (OGC) Web Map Service Implementation Specification (version 1.3.0)	5 (33.33%)	5 (33.33%)
4.	Cadastral Data Content Standard	5 (33.33%)	4 (26.67%)
5.	INCITS 453: 2009, Information technology - North American Profile of ISO 19115:2003 - Geographic information - Metadata (NAP - Metadata)	5 (33.33%)	5 (33.33%)

In its response to the RFI from NIST, the FGDC noted that it uses voluntary consensus standards, as defined in OMB Circular A–119, in lieu of government-unique standards whenever possible. In 2010, the FGDC endorsed 64 external standards, many of which were voluntary consensus standards.

The FGDC emphasized its participation in the InterNational Committee on Information Technology Standards Technical Committee L1 (INCITS L1), Geographic Information Systems; International Organization for Standardization (ISO) Technical Committee 211, Geographic information/ Geomatics; the Open Geospatial Consortium, Inc. (OGC); and the Geospatial-Intelligence Standards Working Group (GWG). By participating in these standards bodies, the

FGDC leverages the knowledge, skills, and talent of the geospatial community in the development of standards that enable information, data, and service interoperability.

The FGDC, in its response to the RFI, identified factors that hinder participation in standardization activities and (or) the adoption and use of standards:

- *ISO and the American National Standards Institute (ANSI) charge a fee for standards.* It is difficult to promote the adoption of standards when State and county agencies and local and Tribal governments are obliged to purchase standards, even if they were involved in their development and (or) review.
- *Voluntary consensus standards often have copyright restrictions that negatively affect their promulgation and use.* Nonetheless, OMB Circular A–119 directs Federal agencies to "observe and protect the rights of the copyright holder and any other similar obligations."
- *Federal agencies have limited resources to commit to standardization activities.* The return on investment for standardization activities is often a long-term outcome. The priority level for standardization activities is reduced when agencies need to show short-term benefits.
- *The Federal civilian community needs better incentives to adopt standards.* The FGDC noted that one disincentive to adopting and using standards is the current lack of a mechanism to enforce the use of standards.

More information on the RFI is available at: *www.nist.gov/ director/sco/index.cfm* (search on "SOS RFI Comments Summary").

FGDC ADVANCES THE NATIONAL SPATIAL DATA INFRASTRUCTURE

The Geospatial Line of Business Advances

Priorities for the Geospatial Line of Business in fiscal year 2011 have been to

1. increase the Federal Government's return on investment for geospatial data and technology,
2. increase the ability of Federal agencies to utilize existing geospatial data from Federal and non-Federal partners,
3. increase the Federal Government's ability to utilize high-value geospatial data for decisionmaking in a timely manner, and (4) provide a platform for the service and use of common geospatial data, applications, and services.

These goals are being met through the continued cross-agency development of a Federal geospatial asset portfolio management capability, as outlined in the "Geospatial Line of Business, OMB Circular A–16 Supplemental Guidance," released by OMB November 10, 2010; and the development of the Geospatial Platform, as defined in the "Modernization Roadmap for the Geospatial Platform," March 2011; and the ongoing drafting of the

"Building the Business Case for the Geospatial Platform, the Value Proposition" document. Many of the 2011 tasks have moved the FGDC toward the Phase I implementation of the Geospatial Platform. The major efforts for the Geospatial Line of Business have been OMB Circular A–16 Supplemental Guidance approval, beginning portfolio management implementation, and planning and initial implementation of the Geospatial Platform.

The Circular A–16 report cover.

These efforts demonstrate forward movement in how we do business for geospatial activities. Below is a summary of each effort and highlights of accomplishments.

Geospatial Line of Business Accomplishments

- The Office of Management and Budget endorsed and released Circular A–16 Supplemental Guidance, November 10, 2010.
- The FGDC Steering Committee endorsed the new list of National Geospatial Data Asset Themes in alignment with A–16 Supplemental Guidance.

Line of Business, OMB Circular A–16 Supplemental Guidance," released by OMB November 10, 2010; and the development of the Geospatial Platform, as defined in the "Modernization Roadmap for the Geospatial Platform," March 2011; and the ongoing drafting of the "Building the Business Case for the Geospatial Platform, the Value Proposition" document. Many of the 2011 tasks have moved the FGDC toward the Phase I implementation of the Geospatial Platform.

The major efforts for the Geospatial Line of Business have been OMB Circular A–16 Supplemental Guidance approval, beginning portfolio management implementation, and planning and initial implementation of the Geospatial Platform.

These efforts demonstrate forward movement in how we do business for geospatial activities. Below is a summary of each effort and highlights of accomplishments.

A–16 Supplemental Guidance Endorsed

The 2011 endorsement of the OMB Circular A–16 Supplemental Guidance document by the Office of Management and Budget provides definition and clarity to selected elements of OMB Circular A–16 to help facilitate the adoption and implementation of a coordinated and effective Federal geospatial asset management capability. This management capability improves support of mission-critical business requirements of the Federal Government and its stakeholders.

The Supplemental Guidance primarily focuses on geospatial data as capital assets and provides the foundation for a portfolio management approach to a National Geospatial Data Asset (NGDA) Portfolio comprised of NGDA Themes and their associated NGDA Datasets.

Portfolio Management Begins Implementation

A key tenant of the Circular A–16 Supplemental Guidance is having an understanding of the inventory of NGDA datasets. The development of the current NGDA inventory affords NGDA Theme Leads the opportunity to review, validate, and evolve those data assets for which their organizations hold stewardship responsibilities. In order to effectively show how the NDGAs support the business requirements of the Federal Government and its partners, the data inventory needs to compare characteristics such as content, quality, application, validity, and definition in order to have a consistent process for determining the data that belongs in the inventory. This initial NGDA inventory activity identified a baseline portfolio in support of an emerging full portfolio management process.

The portfolio management implementation strategy serves as a guide for the actions needed to reach full management of the portfolio of NGDA datasets. A phased approach for achieving full implementation includes continual improvements and annual reporting on progress. Each phase supports the outcomes identified in the OMB Circular A–16 Supplemental Guidance. The portfolio management process consists of six stages: inventory, selection, organization, management, evaluation, and monitoring. This process also includes the setting of Federal geospatial dataset priorities to ensure that NGDA Datasets are available to support the mission needs of the Federal Government and its partners, as determined by Federal agencies and their partners and as recommended to OMB.

The Geospatial Platform Initiative

Geospatial assets are an integral part of government and private day-to-day operations. While the business needs of stakeholders vary, different operations often require the use of similar assets. Coordinating the development and integrated use of the Nation's geospatial assets is a complex effort, but the geospatial community has come together to create an initiative—the Geospatial Platform— to promote improved coordination and more effective use of geospatial information. The Geospatial Platform is designed to leverage geospatial expertise and existing geospatial data, services, applications, and infrastructure from across the government to enhance the Nation's utilization of geospatial resources. The Geospatial Platform is an Internet-based capability providing shared and trusted geospatial data, services, and applications for use by government agencies and partners to meet their mission needs and for use by the public and provides opportunities to view and conduct business in new and innovative ways.

GEOSPATIAL PLATFORM

The Geospatial Platform will function as the service and delivery mechanism for the Federal portfolio of geospatial data, applications, and services, and as the integration point for the discovery, access, and joint use of Federal and non-Federal geospatial data and services. The accomplishments for the year include:

- Developed the "Modernization Roadmap for the Geospatial Platform, Version 4" (released by OMB).
- Developed and launched IdeaScale to solicit comments on the Modernization Roadmap for the Geospatial Platform.
- Released the redesigned Geospatial Platform website (www.geoplatform.gov).
- Members of the Geospatial Platform's Technical Deployment Task Team (TDTT) developed a "Platform" prototype and architecture to frame requirements for future platform offerings.

Geospatial One-Stop Migrating to Data.gov

The Geospatial One-Stop (GOS) portal and catalog continued to be a central resource for inventory of geospatial assets in the United States in support of the Geospatial Line of Business and the National Spatial Data Infrastructure (NSDI). The catalog and data visualization capabilities of GOS are also identified as key components of the technical solutions in the proposed Geospatial Platform Version 0 and Version 1 implementations.

Federal, State, local, academic, and Tribal data and service resources are registered through individual metadata entries. The holdings of the GOS database grew substantially in fiscal year 2011. With more than 678,000 individual metadata records contributed by 450 publishers, this is a 52-percent increase in records from fiscal year 2010. The National Oceanic and Atmospheric Administration, U.S. Census Bureau, and U.S. Geological Survey were the main contributors of these additional records. The significant increase in the number of records in the GOS database continued to be driven by the integration of GOS with Data.gov.

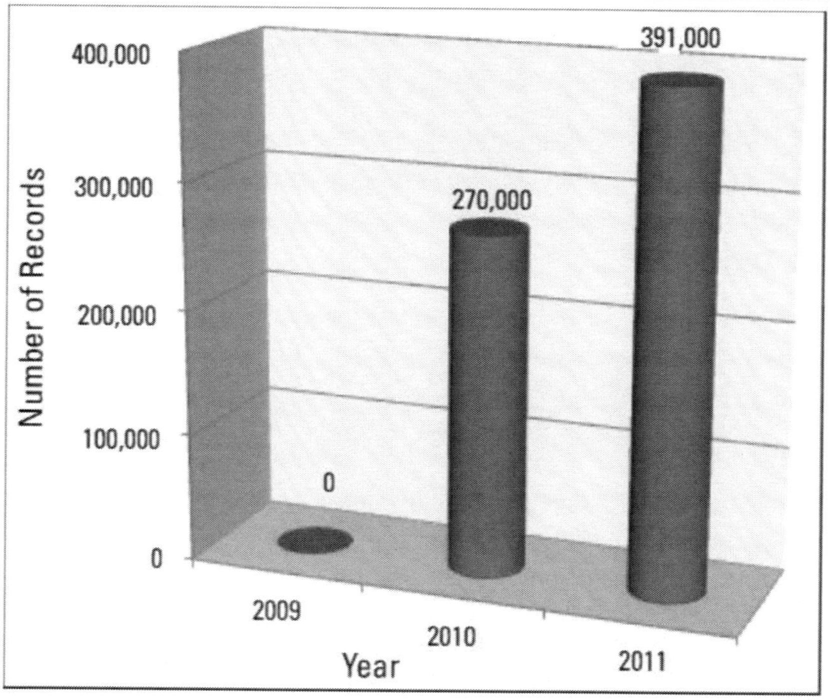

Graphic showing the number of records shared by Geospatial One-Stop (GOS) with Data.gov.

The GOS metadata catalog plays a central role in the population of Data.gov, now contributing both Federal and non-Federal geospatial assets. In all, over 391,000 records are shared from GOS with Data.gov, which means that 90 percent of the records available in Data.gov are geospatial data from GOS. This is a good example of how shared and integrated capabilities can increase governmental transparency and efficiency.

The GOS team continued to work with Federal agencies to promote collections of records through GOS for discovery in Data.gov. Several large collections of metadata were added to GOS/Data.gov, supplementing additions from the U.S. Census Bureau, the U.S. Environmental Protection Agency, and other agencies. These include US Topo (the next generation of digital topographic maps from the U.S. Geological Survey), high-resolution aerial orthoimagery from the U.S. Geological Survey, and Shuttle Radar Topography Mission data from the National GeospatialIntelligence Agency and the National Aeronautics and Space Administration.

The GOS portal began its integration with Data.gov during fiscal year 2011, and this transition is expected to be complete in fiscal year 2012. At that time, Data.gov will become the official means for accessing metadata resources managed in the NSDI Clearinghouse Network. The goal is to create an integrated system to catalog and provide access to geospatial and non-geospatial data and services and to support the existing requirements and objectives of NSDI and the Geospatial Platform. Content has been migrated from the GOS catalog to a new repository integrated with Data.gov at the U.S. General Services Administration. This forms the basis of metadata management and search for the NSDI. The integration task will also provide the initial capabilities required for the Geospatial Platform activity—cataloging, visualization, and access to remote (agency) and common services, and leveraging cloud-based and agency-hosted data and services.

Once the GOS metadata collection is migrated to Data.gov, Data.gov will include more than just the Federal, downloadable datasets that are currently available. Data.gov will utilize the catalog for search and access to Federal validated data, and the Geospatial Platform will utilize the catalog for search and retrieval of Federal and non-Federal records and registered services. Geospatial data visualization capabilities in Data.gov will be enhanced and a geospatial data search capability will be added. In these ways, Data.gov will encompass and extend much of the functionality that was formerly available through the GOS portal.

GeoCloud Sandbox Initiative

The GeoCloud Sandbox Initiative is an FGDC–General Services Administration sponsored activity to develop geospatial application server software packaging that can be used in the Cloud environment. The FGDC Technology and Architecture Working Group sought Federal participants with application server software who are interested in prototyping deployment of public geospatial data and service offerings in a Cloud environment for up to one year.

The objective of the GeoCloud Sandbox Initiative is to provide technical assistance; external hosting monitoring; and reporting on configuration options, results, usage, and costs for mission-critical geospatial data and software within a Federal cloud-computing environment. Key items addressed in fiscal year 2011 were:

- identifying requirements-driven solution architectures (comprehensive software environments) for variously sized deployments of geospatial data and services;
- documentation and cost analysis to support scalability, reliability, and redundancy;
- addressing system security concerns to expedite Federal Information Security Management Act (FISMA) certification and accreditation for agency adoption; and
- supporting and collecting cost comparison information from agencies for existing and externally hosted Cloud solutions.

Accomplishment of the GeoCloud Sandbox Initiative

Amazon Cloud services are being used in the GeoCloud Sandbox Initiative to support both open source and commercial geospatial application servers for nominated FGDC member agency projects. Benefits of the Cloud include scalable capacity, performance, shared system security profiles, reduced time to deploy, and manageability though variable monthly costs of operation. Projects range from small to large and project experience can be shared and applied by other agencies.

Ten projects were identified within the U.S. Census Bureau, U.S. Department of Agriculture, U.S. Environmental Protection Agency, U.S. Fish and Wildlife Service,U.S. Geological Survey, and National Oceanic and Atmospheric Administration. These projects were typically existing, public-facing Web service environments that included standard geospatial software for access and (or) analysis. By starting with existing systems, a fair comparison of in-house and cloud-hosting costs and procedures can be made. Through the redeployment of certified systems, the system security approval process can be accelerated. By sharing common packages of software, agency projects can share the "platform" certification burden and accelerate deployment times, adding in only their data and minor customization.

Two geospatial "platform as a service" environments (also known as reference platforms) were developed based on the requirements of the participating projects. Common software and operating systems were identified to create the reference platforms in order to maximize re-use. An open-source reference platform includes Linux (CentOS), Apache, MySQL, PostgreSQL, Java, Tomcat, OpenGeo's GeoServer, and GeoNetwork products. The commercial reference platform includes Windows 2008 Server R2, IIS, Java, Tomcat, and Esri ArcGIS Server, and related programming environments. Installation scripts were written using the Amazon CloudFront technology that enables projects to rapidly configure and deploy virtual machines in Amazon. Periodic assessments of dependencies between the software packages were made, resulting in adjustments to the versions and installation scripts.

The U.S. Census Bureau deployed its 2009 TIGER Line file server using the Linux CentOS image, offering an enhanced ftp/http data access service for all public access to 2009 TIGER data. In the Windows environment, the key system deployed was the Wetlands Mapper Application using the ArcGIS Server, developed by the U.S. Fish and Wildlife Service and deployed by the U.S. Geological Survey information technology staff. Performance and reliability of the deployed services in both cases was comparable or superior to the legacy service environment in the agency data centers. Additional configuration capabilities exist in the Cloud environment for load balancing and to automatically scale additional service capacity in times of high demand.

The most challenging aspect of migration to the Cloud has been acquiring agency security approval for the Cloud applications. In most cases, the existing agency system security approval (Certification and Accreditation - C&A) was used as a basis for a Cloud C&A, but the hosting infrastructure, interconnect agreements, and service-level agreements desired by the agencies were sometimes difficult to document or acquire in order to issue an Authority to Operate (ATO). These infrastructure-based security issues are being addressed through GeoCloud use of the recently GSA-approved Infrastructure as a Service (IaaS) Blanket Purchase Agreement for Cloud services, acquired through Apps.gov. This has provided a Federal-wide security approval for the IaaS solutions that agencies can accept and incorporate into their systems-accreditation process.

National Geospatial Advisory Committee Promotes Communication

The National Geospatial Advisory Committee (NGAC) is a Federal advisory committee sponsored by the Department of the Interior to provide external advice and recommendations to the member agencies of the FGDC. The NGAC includes 28 committee members representing a variety of organizations involved in geospatial issues, at all levels of government, the private sector, non-profit organizations, and academia. The list of members is available at *www.fgdc.gov/ngac*. The NGAC meets three to four times per year and has established subcommittees that conduct research and develop draft products between committee meetings. The NGAC promotes two-way communication on issues of common interest to the national geospatial community and provides a forum to convey views representative of our partners and stakeholders. The NGAC has served as a valuable source of advice and feedback for the FGDC. Over the past year, the NGAC has analyzed and provided recommendations on a number of key geospatial policy issues. Highlights of the NGAC's 2011 activities include the following:

- The NGAC developed recommendations on several important topics in 2011, including the Geospatial Platform Initiative, *The National Map*, geolocation privacy, the National Land Imaging Program, Transportation for the Nation, and the Federal Aviation Administration's laser pointer issue, which could impact the collection of LiDAR data.
- The NGAC adopted an Action Plan for committee activities in 2011/2012, including goals and subcommittees organized under the following areas: - Innovative Strategies for Geospatial Programs and Partnerships
  - Geospatial Platform
  - Geospatial Workforce Development
  - Emerging Technologies
  - Geospatial Partnerships with Tribal Governments
- The NGAC also designated Points of Contact (POCs) for significant Federal geospatial programs and initiatives. The role of the POCs is to monitor program activities and keep NGAC members informed of key developments.

The NGAC engaged in a dialogue with FGDC Chair Anne Castle regarding the White House budget initiative for the National Land Imaging Program and adopted the following resolution:

> "The NGAC strongly supports the establishment of the National Land Imaging Program as described in the Administration's fiscal year 2012 budget proposal and we encourage the Department of the Interior to implement the program as soon as possible. NGAC looks forward to providing advice to the Department on the development and implementation of the program. This recommendation is made in the context of a more

comprehensive multi-agency, multi-resolution, multi-sensor imagery program for the nation."

- The NGAC adopted a paper on local government GIS best practices.
- The NGAC adopted a paper on best practices for interagency data sharing.
- The NGAC held a series of "spotlight sessions" that brought together experts on key topics to discuss new approaches and identify best practices and lessons learned. The spotlight sessions focused on the topics of interagency data sharing, geospatial workforce development, and national land parcel data.

In 2012, the NGAC will continue to provide advice and feedback on the development and implementation of key geospatial topics, including innovative strategies for geospatial programs and partnerships, the Geospatial Platform, geospatial workforce development, emerging technologies, and partnerships with Tribal governments.

These multifaceted activities will be a major focus of the NGAC's work over the coming year.

NGAC Leadership Transition

In February 2011, Secretary of the Interior Ken Salazar appointed a new Chair and Vice Chair to lead the NGAC. Dr. David Cowen, Professor Emeritus of Geography at the University of South Carolina, was appointed as NGAC Chair. Dr. Jerry Johnston, the Geospatial Information Officer at the U.S. Environmental Protection Agency, was appointed as Vice Chair. Dr. Cowen and Dr. Johnston are well-known and respected leaders in the geospatial community, and they bring a wealth of experience and expertise to their new leadership roles. The new leadership team replaces the outgoing Chair and Vice Chair, Anne Hale Miglarese of Booz Allen Hamilton and Steven Wallach, recently retired from the National Geospatial-Intelligence Agency. Ms. Miglarese and Mr. Wallach served with distinction as the first leadership team for the NGAC to establish the NGAC as a trusted voice for the geospatial community.

Standards Progressing Forward

United States Thoroughfare, Landmark, and Postal Address Data Standard

In February 2011, the FGDC Steering Committee endorsed the United States Thoroughfare, Landmark, and Postal Address Data Standard ("Address Data Standard"). The standard contains four parts: address data content, classification, transfer, and quality.

Street addresses, which are covered by the Address Data Standard, are the location identifiers most commonly used by various levels of government and the public. Street addresses are critical for administrative records, emergency and first response, research, marketing, mapping, geospatial information systems, routing and navigation, and law enforcement. The standard provides Federal, State, local, and Tribal agencies with a means to share address information, as well as a methodology for improving response to emergency and non-emergency service requirements, with associated improvements in efficiency, effectiveness, and economy.

"The United States Thoroughfare, Landmark, and Postal Address Data Standard will have a significant impact at all levels of government and is an essential component of the National Spatial Data Infrastructure," said Ivan DeLoatch, Executive Director of the FGDC. "We encourage government agencies, as well as others, to implement the standard for it provides a foundation for understanding and developing solutions for the many challenges in our communities."

The U.S. Postal Service (USPS) and the National Emergency Number Association (NENA) were involved in the development of profiles for the Address Data Standard. The "Postal Addressing Profile of the Federal Geographic Data Committee United States Thoroughfare, Landmark, and Postal Address Data Standard" outlines the equivalencies and differences between the Address Data Standard and USPS Publication 28, "Postal Addressing Standards."

Similarly, the "Profile Reconciling the FGDC United States Thoroughfare, Landmark, and Postal Address Data Standard" and the NENA's "Next Generation 9-1-1 (NG9-1-1) Civic Location Data Exchange Format (CLDXF) Standard" outline equivalencies and differences between the two standards. The FGDC awarded Spatial Focus, Inc. a 2011 National Spatial Data Infrastructure Cooperative Agreements Program (NSDI CAP) grant for prototyping data quality and data exchange tools to assist government agencies in implementation of the Address Data Standard.

Coastal and Marine Ecological Classification Standard

Coastal and marine planners and managers are faced with complex challenges when making decisions about habitat conservation and resource management. Given this complexity, and recognizing the vastness of the marine environment for which the United States has jurisdiction, the requirement for a standard national classification for coastal and marine systems was identified.

To fulfill this requirement, the National Oceanic and Atmospheric Administration and NatureServe developed the Coastal and Marine Ecological Classification Standard (CMECS). CMECS is relevant to all U.S. coastal and marine environments and can be applied locally, regionally, nationally, and even globally.

A standardized approach enables coastal and marine planners and managers to develop strategies for resource management and conservation for all of North America's coasts and oceans.

CMECS was released for public and peer review in 2010. The FGDC awarded a 2011 NSDI CAP grant to NatureServe to support resolution of comments from public and peer review and production of a final draft. A final draft CMECS is expected in December 2011.

Return on Geospatial Investment:
Iowa LiDAR Project

The State of Iowa is seeing many benefits and savings from its LiDAR elevation project and from freely sharing the data with city, county, State, and Federal agencies and private engineering firms.

Business benefits and cost savings have been seen in numerous areas, including reducing the cost of planning topographic surveys for construction projects, county planning for wind farm and industrial siting, city water and sewer improvement projects, and emergency and disaster management. These benefits and savings are backed up by a recent Return on Investment study completed by the Iowa Geographic Information Council and utilized components of the FGDC Return on Investment methodology developed through the Cooperative Agreements Program. The study showed an estimated benefit of $5 million per year.

Here are several examples of how free LiDAR data provides benefits and saves time and money.

- The city of Leon in Decatur County took advantage of the availability of the data to put in place a plan for a sewer improvement project. Since the project was deemed "shovel ready" the city received about $2 million in American Recovery and Reinvestment Act funding for the improvement project.
- Private engineering firms using the free LiDAR data are able to save small cities several thousand dollars per project and reduce the project times by 50 percent.
- County governments are able to reduce the cost of their aerial photography projects by as much as $36 per square mile, or about $21,000 for a typical county. The LiDAR data also improves the accuracy and quality of the photos.
- Using LiDAR data, aerial photos, and other data, the U.S. Army Corps of Engineers was able to determine that if a potential breach of the Iowa River near Marshalltown occurred, it would most likely flow in such a manner as to not be of danger or risk to the Sutherland Power Plant. This analysis determined that a $500,000 streambed enhancement project would not be necessary at that time, saving money for the local government and taxpayers.
- Iowa's LiDAR data was also recently used to recalculate flood inundation by the Missouri River in western Iowa.

The study is recognized as being conservative in its estimates—recent Web statistics indicate that one terabyte of the LiDAR data is being downloaded freely every week by a variety of academic, government, and business users. See *www.geotree.uni.edu/lidar Project.aspx.*

These examples show how data partnering and sharing provide positive, measureable cost savings and cost avoidance.

Federal Trails Data Standard

In 2011, the FGDC Steering Committee endorsed the Federal Trails Data Standard for FGDC endorsement. Governmental agencies and the public recognize trails as important recreational and cultural resource corridors. The National Park Service, the Bureau of Land Management, U.S. Fish and Wildlife Service, and U.S. Forest Service have worked with each other and with States, local governments, and trail organizations for many years to promote and develop trails for the benefit of the public.

Trail information is a necessity throughout the trails management lifecycle, from planning through design, construction, operation, and maintenance. The U.S. National Park Service, Bureau of Land Management, U.S. Fish and Wildlife Service, and U.S. Forest Service developed the Federal Trails Data Standard to facilitate the access, exchange, and use of trail information and increase the capability for enhanced and consistent trails mapping, inventory, monitoring, condition assessment, maintenance, costing, budgeting, information retrieval, and summary reporting.

Geospatial-Intelligence Standards Working Group

The FGDC recognizes the need for common standards across the defense, intelligence, and civilian communities, as these communities establish a common operating environment. The FGDC is a member of the GeospatialIntelligence Standards Working Group (GWG) chaired by FGDC member agency the National Geospatial-Intelligence Agency. The GWG provides a consensus-based community forum authorized to prescribe and mandate geospatial standards for use by Federal Government organizations within the Department of Defense and the intelligence community. The FGDC is the GWG's link to the civilian community. In 2011, the FGDC worked to develop a standards review and approval process to align with the process used by the GWG so that there will be common standards among the defense, intelligence, and civilian communities.

Geospatial Metadata: Workhorse of the NSDI

Metadata continues to be a fundamental component of NSDI initiatives across the board. Metadata records published though the Geospatial One-Stop (GOS) portal comprise the majority of Federal records within Data.gov.

As the transition from GOS to Data.gov as a catalog for the Geospatial Platform progresses, the FGDC will continue to provide a venue for non-Federal records in support of data sharing among the broader NSDI community within an integrated catalog shared with Data.gov. Data sharing is increasingly recognized as a GIS best practice by NSDI stakeholders, and much of the 2011 effort was focused on supporting the National Science Foundation (NSF) DataONE data sharing effort and the National States Geographic Information Council (NSGIC) Data Sharing Working Group.

Metadata is also fundamental to data development in compliance with Framework and other spatial data standards. The NSDI Cooperative Agreements Program (NSDI CAP) sponsored the development of data and metadata that demonstrate the application of the recently approved United States Thoroughfare, Landmark, and Postal Address Data Standard. Products from the NSDI CAP project will serve as examples for use by NSDI stakeholders as they implement the standard.

With an eye to the future, focus is also directed toward the challenges, tools, and processes for migrating from the FGDC's Content Standard for Digital Geospatial Metadata to the International Organization for Standardization (ISO) Geographic Information – Metadata standard. Planning is underway for a Metadata Summit that will bring together a group of metadata subject matter experts to identify challenges and opportunities presented by the new metadata standard and to begin identifying strategies for effective implementation.

Cooperative Agreements Program Builds the NSDI

To encourage and enable building the NSDI at all levels of the geospatial data community, the FGDC sponsors the Cooperative Agreements Program (CAP). The benefits to the community are not only financial but more importantly validate an organization's geospatial endeavors, which leads to new opportunities. The CAP has created collaborations within all sectors of government, helped develop an understanding of geospatial information in organizations new to the NSDI, provided seed money to enable geospatial organizations to participate in the national effort to implement the NSDI, promoted the importance of geospatial data standards, promoted the development of standardized metadata in hundreds of organizations, and greatly expanded implementation of geospatial services on the Internet.

Feedback on the NSDI CAP

Montana: "[The program] improves the quality of data available at the national level by facilitating the movement of local authoritative data to the State, then to the national level. It also helped to foster relationships with local data providers and emphasize the importance of data sharing which in turn benefits the Federal level."

The program's diversity can be seen in the 27 CAP projects completed this year. The GeoCommons platform was extended to enable it to dynamically georeference Federal Government tabular data, which will allow data to be discovered, accessed, and applied in its geographic context; Geocommons.com. Four projects developed client/server-mediated applications that can access, search, and exploit the geospatial data and services referenced by the Geospatial One-Stop portal to support specific geographic or discipline requirements. For more information see *www.fgdc.gov/grants/2009CAP/ 2009CAPDescriptions#category2*. Illinois, Indiana, and Missouri created or enhanced numerous partnerships between regional and local governments and demonstrated viable solutions to solve data interoperability problems. Projects in Alaska, Michigan, Montana, and Utah focused on building data stewardship for transportation geospatial information. Complete guides to the National Vegetation Classification Standard (Version 2) were developed for online and classroom use; *mtnhp.org/ecology/nvcs/*. Finally, the CAP continued to sponsor projects that serve to train a variety of interest groups in the value of metadata and in its implementation and sharing. This year over 141 people were introduced to metadata concepts though a variety of workshops and online and in-person presentations. In the 2011 program, CAP awards were given to 20 projects that address six categories. The planned CAP budget for fiscal year 2012 is just over $800,000, and these funds will be used to support up to 41 projects. For more information, please see *www.fgdc.gov/grants.*

Fifty States Initiative Promotes Coordination

This year marked the sixth anniversary of the Fifty States Initiative. The Initiative strengthens statewide coordination by advancing strategic and business plans and helps identify activities where strategy for Federal partnerships can and should occur. Since 2006, 48 States plus the District of Columbia and the U.S. Virgin Islands have received support through the NSDI CAP program. In 2011, eight States (Delaware, Hawaii, Idaho, Kentucky, Maine, Michigan, Mississippi, and Oregon) plus the U.S. Virgin Islands completed Initiative projects and ten awards were given to projects in California, Florida, Idaho-Montana (joint project), Louisiana, Minnesota, New Jersey, Tennessee, Utah, West Virginia, and Wisconsin. This year's "Measuring Progress of the Fifty States Initiative" status report noted that NSDI CAP helps to formalize efforts toward and improvement of statewide coordination, communication, and collaboration among stakeholders. In addition, States that have plans in place were in a better position to compete for American Reinvestment and Recovery Act (ARRA) funds that furthered their geospatial capabilities and their contributions to national level data. As the Fifty States Initiative continues into its seventh year, the emphasis will be on furthering business planning and the plan implementation process that helps States forge key partnerships. For more information about the Initiative, please see www.fgdc.gov/ policyandplanning/ 50states.

International Activities Progressing

The FGDC Secretariat continues to support international efforts through the Global Spatial Data Infrastructure (GSDI) Association and the intergovernmental Group on Earth Observations. These activities are underpinned by the adoption of common techniques and standards to promote interoperability worldwide and thereby facilitate access to geospatial data and services. The FGDC supports the GSDI Association efforts to increase international collaboration on spatial data infrastructure (SDI) development. In support of this, the FGDC coordinates and supports the GSDI Small Grants Program that promotes SDI adoption in developing countries around the world.

In 2011, fourteen projects were supported. The GISCorps of Urban and Regional Information Systems Association partners in the GSDI Small Grants Program and provides professional services in support of this effort.

Through the GSDI Association, FGDC funds the publication of regional SDI newsletters. In addition, the FGDC hosts a visiting scientist from Korea who supports the Asia Pacific regional newsletter and other GSDI activities.

The FGDC Secretariat continued support for the intergovernmental Group on Earth Observations (GEO) and its Global Earth Observation System of Systems (GEOSS). In fiscal year 2011 the Secretariat supported the revision of the 2012–2015 workplan and committee structure.

Ivan DeLoatch, FGDC Secretariat Executive Director, is a co-chair for the Architecture and Data Committee, and Doug Nebert, FGDC Secretariat Senior Advisor for Geospatial Technology, is Task Lead for the key architectural task (AR-09-01A) within GEO. The FGDC and U.S. Geological Survey developed and continue to operate and evolve core catalog service capabilities for GEOSS—the GEOSS Clearinghouse and Component and Service Registry— through technical support agreements with George Mason University. Significant enhancements to the capability of GEOSS were developed for presentation to the GEO Plenary Session in November 2011.

NSDI Training Expands Beyond Metadata

FGDC continues to work with stakeholders to develop a range of NSDI-related training materials. A cooperative agreement with the University of Wyoming resulted in the development of a full suite of NSDI Framework Data training materials that go beyond basic presentation slides by including links to resources, an audio track, quizzes, and other dynamic features. The materials are available from the NSDI Online Training Program website (www.fgdc.gov/ training/nsdi-training-program/online-lessons#lessons).

GeoCircle is a new website that facilitates communication about geospatial issues. It allows Native American Tribes to connect with the Department of the Interior (DOI). GeoCircle allows users to share information about:

- Tribal points of contact for geospatial activities
- DOI liaisons to Tribes for geospatial issues
- Grant and funding resources
- Training resources
- Enterprise license agreements
- Conference passes
- Documents contributed by both the Tribes and DOI
- Dynamic content contributed by Tribes and DOI

If you would like to participate in building and using GeoCircle, please take the first step by registering. If you register, you will be able to read and contribute content to the GeoCircle website. For instructions on how to register see *www.fgdc.gov/geocircle/ GeoCircle_HowToRegister_ v3.pdf.*

The FGDC is proud to support GeoCircle and provides information on FGDC activities and opportunities of interest to the GeoCircle community.

FGDC supports the National Science Foundation (NSF) DataONE program efforts to develop training materials that encourage and guide scientists in best practices for sharing research data within their scientific communities and with the public. The materials are being developed through the DataONE Community Engagement and Education Working Group and will be available from the DataONE website (*www.dataone.org*) once completed.

Metadata remains the most commonly requested NSDI training topic. FGDC partnered with the U.S. Geological Survey National Biological Information Infrastructure (NBII) program to lead a metadata "Train the Trainer" workshop for the Bureau of Land Management. FGDC also partnered with the National Oceanic and Atmospheric Administration's National Coastal Data Development Center (NCDDC) to provide introductory metadata training to coastal partners in State and local governments and their private-sector associates. Through the Cooperative Agreements Program, FGDC supported university, Tribal, and private-sector stakeholders in developing and delivering metadata training workshops for State, local, and Tribal communities.

FGDC Subcommittee and Working Group Reports

The FGDC currently has nine subcommittees that form communities of interest for nine data themes.

In addition, four FGDC working groups crosscut the subcommittees and focus on infrastructure issues common to or supporting many of the NSDI data themes.

Federal agencies lead and have responsibility for the thematic subcommittees and working groups. Below are the subcommittee and working group high-level summaries of this year's activities. Additional information is available at www.fgdc.gov/ participation.

Cadastral Subcommittee

- The FGDC Subcommittee for Cadastral Data continued to provide updates to the existing national inventory (*www.gisinventory.net*) of cadastral data. Almost all the States now have an identified State Cadastral Coordinator and many States have assumed responsibility for sustaining the inventory of local contacts.
- Updated sets of parcel data were provided to the wildland fire community, and efforts to establish sustainable standardized parcel datasets at the State level built on last year's efforts. The States of Arizona, California, New Mexico, Nevada, and Washington made significant advances toward a sustainable system.
- The development of standardized Public Land Survey System (PLSS) datasets has been completed and updated in nine western States, and this data is being provided to the States as requested.
- The Federal Parcel Work Group completed the first draft of their report on the inventory of Federal and management agencies and a proposed core data standard for Federal land parcels.

Cultural and Demographic Statistics Subcommittee

- The United States Thoroughfare, Landmark, and Postal Address Data Standard was completed and adopted as an FGDC standard in February 2011.

Geodetic Control Subcommittee

- In fiscal year 2011, the subcommittee continued to engage its partners in its planning and preparation for the new geometric and geopotential datums. Two key subcommittee meetings were held that brought together representatives from 17 Federal agencies, 3 State offices, several private companies, and the National Society of Professional Surveyors. The overarching focus of those meetings was to plan and prepare for the datums of the future that will shift our Nation from a National Spatial Reference System (NSRS), defined by passive monuments, to a system based on observations of real-time Global Navigation Satellite Systems and measurement of the earth's gravity field. Those changes will result in (1) profound improvements to the accuracy of the NSRS, and (2) enhanced access to the NSRS.

Geologic Subcommittee

- On behalf of the FGDC, the Digital Cartographic Standard for geologic maps continues to be supported by the U.S. Geological Survey National Geologic Map Database project. Collaboration with Esri has resulted in the release and subsequent update of a subset of the FGDC symbols, created as Cartographic Representations for use in ArcGIS. The Subcommittee is now planning for a significant update to the standard (pending available funding), to begin in late 2011. Guidance on the plan was sought from technical experts at the annual Digital Mapping Techniques (DMT) meeting in May 2011.
- A new database design for publication of geologic maps ("NCGMP09") was proposed by the U.S. Geological Survey; the Subcommittee has begun evaluation of this and other designs, targeted at a Federal standard.

Marine and Coastal Spatial Data Subcommittee

- Member agencies of the Marine and Coastal Spatial Data Subcommittee were active participants in the development of the Coastal and Marine Spatial Planning National Information Management System (NIMS). This effort, led by staff from the National Ocean Council, included participation by 15 Federal agencies.
- Deployed phase two of the Multipurpose Marine Cadastre (MMC), a Web-based geospatial platform, that provides the baseline information needed for coastal and marine spatial planning efforts, particularly those that involve finding the best location for renewable energy projects and their permit review process. The Bureau of Ocean Energy Management was the first bureau to provide data content directly to the MMC via Web services.
- Subcommittee members continued to develop the Coastal and Marine Ecological Classification Standard (CMECS) and guide it through the FGDC Standards Approval Process.

National Digital Orthoimagery Program (NDOP) Subcommittee

The NDOP Subcommittee is responsible for developing a national strategy for acquisition of orthoimagery data for Federal agencies while creating and utilizing partnerships with State, local, Tribal, and private organizations.

- Details for the two major NDOP programs, the National Agriculture Imagery Program (NAIP) and the Urban Areas Program, are contributed to by the U.S. Department of Agriculture and the U.S. Geological Survey, respectively, and are described on page 36–38.
- Projects conducted by the Technical Management Subgroup include improving color balancing on mosaicked images, horizontal and vertical accuracy testing, evaluating compression ratios, pan-sharpening techniques, and ground control point collection.

Spatial Water Data Subcommittee

- *The Subcommittee for Spatial Water Data, a subcommittee of both the Advisory Committee for Water Information and the FGDC Spatial Water Data Committee, has been focused on implementing water Web services in the hydrologic community using a services-oriented architecture (SOA) for water information. The National Science Foundation's funding of the Consortium of Universities for the Advancement of Hydrologic Science's (CUAHSI) research has produced a framework that supports national water observations data synthesis. The Subcommittee has been examining this SOA approach for common use among the Federal water agencies, realizing that this significant SOA framework goes hand in hand with FGDC's major efforts on the Geospatial Platform.*

Transportation Subcommittee

- A revised Transportation Charter was presented to FGDC to reinstate the subcommittee. The charter was drafted by an internal working group comprised of representatives from the U.S. Department of Transportation (DOT), the U.S. Census Bureau, and the U.S. Geological Survey.
- An open meeting of the subcommittee was held at the Esri Federal User Conference. The meeting had representation from the DOT, the U.S. Census Bureau, the U.S. Geological Survey, and the private sector.

Vegetation Subcommittee

- Developed the National Vegetation Classification Standard Logo.
- Agreed on the development of keys for the National Vegetation Classification.
- Met with the Ecological Society of America Vegetation Panel and established a naming peer review process.
- Revised the National Vegetation Classification Standard Implementation Plan.

Transportation for the Nation Initiative

The U.S. Department of Transportation (DOT), in partnership with the National States Geographic Information Council (NSGIC), sponsored a strategic planning effort for the Transportation for the Nation (TFTN) initiative.

TFTN addresses the goal of coordinating, developing, and distributing nationally significant transportation data, with particular emphasis on road centerlines. The initial strategic planning effort included reaching out to stakeholders from all levels of government and the private sector to gather their input on how the TFTN initiative should proceed. Based on this outreach, the plan recommends leveraging the work already being done at the State DOTs for existing Federal DOT programs. For more information, visit *www.tftn.org*.

Wetlands Subcommittee

- Conducted outreach under the Implementation Plan for the Wetlands Mapping Standard through support of the Association of State Wetlands Managers and the Wetlands Mapping Consortium.
- Published a new online wetland mapping course.
- Conducted a multi-organization Working Group maintenance review of the Wetlands Classification Standard.

The Architecture and Technology Working Group

- *Focused on support for the Technical Deployment Task Team to prototype a Geospatial Platform proof-ofconcept, and the management of the Geospatial Cloud Sandbox Initiative, known as GeoCloud. For more information see page 12.*

Metadata Working Group

- Worked collectively to explore International Standards Organization (ISO) geospatial metadata standards implementation.
- Through quarterly teleconferences and email discussions, the members of the working group have shared information about the status of the standards, issues addressed when applying the standards, software tools developed to support conversion from one standard to another, development of training materials and guidance documents, available metadata editors, and ideas for transitioning from the current FGDC metadata standard (Content Standard for Digital Geospatial Metadata, or CSDGM) to the ISO format.

FGDC Standards Working Group

- The Working Group recommended FGDC endorsement of the United States Thoroughfare, Landmark, and Postal Address Data Standard and the Federal Trails Data Standard. The FGDC endorsed the standards in February 2011.
- Working Group members served on NSDI Cooperative Agreements Program review panels to recommend proposals for standards development or implementation.

User/Historical Data Working Group

- Began a program of outreach to perspective members. To this end, a listserv was set up through the Library of Congress for Working Group members and those interested in its work (*sun8.loc.gov/listarch/uhdwg.html).*
- Began monthly Web and phone meetings for members to discuss issues. Various projects have been generated from these meetings, including:
  - GeoPlatform Version 3 Users/Historical Data Working Group Collected Comments,
  - FGDC Users/Historical Data Working Group Information Sheet, and
  - Survey of Geospatial Data Managers and Databases in the Federal Government (ongoing).

FGDC Goals for Fiscal Year 2012

Geospatial Line of Business

- Develop Office of Management and Budget (OMB) Circular A–16 Supplemental Guidance implementation plans that support portfolio management.
- Implement roles and responsibilities identified in the Supplemental Guidance that support adoption of National Geospatial Digital Asset (NGDA) portfolio management.
- Develop and host targeted trainings around Supplemental Guidance implementation that support the vision for Federal NGDA portfolio management.

Challenges: Obtaining continued participation from Federal partner agencies.

Portfolio Management

- Implement the portfolio management roles and responsibilities of the OMB Circular A–16 Supplemental Guidance.
- Refine the inventory of the NGDA datasets and themes.
- Identify potential automated systems to support performance measurement for the NGDA dataset and themes that can be offered through the Geospatial Platform.

Challenges: Identifying stewardship responsibility and theme ownership and developing meaningful and consistent performance metrics for datasets and themes.

Geospatial Platform

- Develop an implementation plan for the outreach/communication
- strategy in support of the Geospatial Platform.
- Deploy Version 1.0 capabilities for the Geospatial Platform in the Cloud environment in concert with Data.gov.
- Identify and launch pilot geospatial services and (or) applications on *www.GeoPlatform.gov*.
- Develop high-level requirements for offering infrastructure, data, and software as services on the Geospatial Platform.

Challenges: Continuing support from Federal partner agencies, identification of a managing partner for Geospatial Platform operations, and identifying a share investment strategy to support the effort.

Geospatial One-Stop and Data.gov

- The primary goal in 2012 will be completing the migration or integration of the Geospatial One-Stop catalog and capabilities into the Data.gov environment hosted by the U.S General Services Administration.

Challenges: Continuity of capabilities from the legacy environment into the new environment, seamless integration of geospatial and nongeospatial data, and deployment of an intuitive data access interface.

GeoCloud Sandbox Initiative

- In fiscal year 2012, the GeoCloud Sandbox Initiative seeks to continue deployment of the open source and commercial Cloud images for a new set of projects with a targeted focus on common services and data services for the Geospatial Platform and OMB Circular A–16 data.

Challenges: Securing information technology system security approvals for Cloud solutions within existing agency processes and allowing the purchase of usage-based Cloud services that vary monthly like a utility bill.

National Geospatial Advisory Committee

- The National Geospatial Advisory Committee (NGAC) will hold three to four public meetings in fiscal year 2012.
- The NGAC will provide ongoing review, feedback, and recommendations regarding the development and implementation of key issues and initiatives, including innovative strategies for geospatial programs and partnerships, the Geospatial Platform, geospatial workforce development, emerging technologies, and partnerships with Tribal governments. These areas will be a primary focus of the NGAC's work.
- The FGDC will review and respond to advice and recommendations from NGAC.
- The FGDC will complete the next cycle of NGAC nominations and appointments.

Challenges: A key challenge (and opportunity) for the NGAC will be to develop advice and recommendations that identify opportunities to develop partnerships and leverage scarce resources to continue to develop interagency and intergovernmental geospatial initiatives in an environment of declining resources.

Standards

- The Coastal and Marine Ecological Classification Standard (CMECS) will go forward for FGDC endorsement.

- Additional external standards from ISO, ANSI, OGC, and other sources will be submitted for FGDC recognition in 2012. In this regard, the FGDC will leverage the resources of the GeospatialIntelligence Standards Working Group to identify and recommend standards from ISO, ANSI, OGC, and other sources for FGDC recognition and use by FGDC member agencies.

Challenges: The FGDC standards activity needs more active participation, particularly from data-producing agencies represented on the FGDC Executive Committee. The FGDC needs additional member agency participation on focus groups within the Geospatial Intelligence Standards Working Group (GWG) to evaluate standards under consideration by the GWG for adoption and use by the civilian community.

Metadata

- Initiate the National Spatial Data Infrastructure stakeholder transition from the Content Standard for Digital Geospatial Metadata (CSDGM) to the ISO suite of geospatial metadata standards.
- Increase the number and type of online resources available from the FGDC Geospatial Metadata website and organize the information so that resources are more easily accessed.

Challenges: Development of and access to the training, guidance, and tool resources necessary to support ISO metadata implementation; coordinating staff time to develop, coordinate with others to develop, and post online training for the resources; expanding the number and quality of Federal agency metadata records; and demonstrating that metadata development is the "context" component of all data and should be integrated into the data development and management process.

Fifty States Initiative

- In fiscal year 2012, eight NSDI Cooperative Agreements Program awards are planned for the Fifty States Initiative, with a kickoff meeting in spring 2012. The initiative will continue to support the Geospatial Platform.

International Activities

- The Global Spatial Data Infrastructure (GSDI) Small Grants Program will award up to eighteen projects that will support the advancement of spatial data infrastructures in developing countries around the world.
- In terms of goals for the Group on Earth Observations/Global Earth Observation System of Systems (GEO/GEOSS), a current focus is on simplifying access to quality Earth observation data and services. This will be achieved through

enhancements to the existing portal and search components, two of which are hosted and maintained by the FGDC/USGS.

Challenges: Limited registration of datasets that support critical Executive Order 12906 parameters, a variety of data access and distribution policies worldwide, and a need for expanded outreach to promote use of GEOSS to find and exploit Executive Order data.

NSDI Training

- Enhance the Web presentation of materials by utilizing Online Learning System technology and organizing the formation so that resources are more easily accessed.
- Expand the number and type of online training materials available to NSDI stakeholders.

Challenges: Enlisting and supporting participation from subject matter experts for NSDI initiatives.

FGDC Subcommittees and Working Groups

Cadastral Subcommittee

- The Cadastral Subcommittee's Federal Parcel Work Group will develop a draft core data standard for Federal parcel data and develop a strategy to implement it.

Cultural and Demographic Statistics Subcommittee

- United States Thoroughfare, Landmark, and Postal Address Data Standard — Completion of profiles of the standard that will work with the U.S. Postal Service's Publication 28 and the National Emergency Numbering Association's Next Generation 9-1-1 Civic Location Data Exchange Format Standard.

Geodetic Control Subcommittee

- Continue education, outreach, development of transition tools and applications, and capacity-building activities to prepare users for the transition to new geometric and geopotential datums. These activities will converge at the National Geodetic Survey's second Geospatial Summit, which will be held in San Diego, California, in July 2012.
- Identify common objectives and find opportunities for cooperative projects and tasks related to standardization and updates to vertical datums.
- Within each Federal agency, identify products and services that will be impacted by new vertical datums.
- Update existing guidelines, standards, and specifications to support the transition to new vertical datums.

Geologic Subcommittee

- Coordinate with various agencies in the United States and Canada to create an Esri "style file" implementation of the FGDC Geologic Map Symbolization standard. Work with Esri regarding revisions to their Cartographic Representations implementation of the Standard.
- The Standard was published in 2006; since then, suggestions for revisions have been received and compiled. In 2012, pending available funding, begin the first update to the Standard. Release revised sections to the Standard as appropriate.
- Evaluate the "NCGMP09" database design for publication of geologic maps that was proposed by the U.S. Geological Survey. Evaluate database designs used by other agencies. Work toward consensus on an overall Federal standard design, or standard database elements (such as common science terms and database field names).

Marine and Coastal Spatial Data Subcommittee

- Release of the Coastal and Marine Spatial Planning National Information Management System (NIMS) Prototype Portal.
- FGDC adoption of the Coastal and Marine Ecological Classification Standard (CMECS).

National Digital Orthoimagery Program (NDOP) Subcommittee

- Future NDOP plans include devising a new 3-year NAIP partnership agreement. All agencies that use NAIP imagery are encouraged to contribute to the plan.

Spatial Water Data Subcommittee

- The Subcommittee, mindful of the framework for services-oriented architecture, will focus on the national delivery of water quality and quantity results using standard geography, GIS, and internationally adopted standards to harmonize the delivery of sampling points and time series data on flow, water levels, water quantity, and water quality.
- The Subcommittee for Spatial Water Data will begin contacting Federal water agencies to work with them to assess what would be the advantages, disadvantages, and level of effort involved in implementing this services-oriented architecture.

Transportation Subcommittee

- Identify a baseline of transportation datasets (non-road) that would complete Transportation for the Nation.
- Identify issues and solutions on how to utilize transportation geospatial data that would contribute to livability analyses in the coming years.

Vegetation Subcommittee

- Host a field trip and organized session for the 2012 Ecological Society of America annual meeting.
- Develop a training workshop related to the use of the National Vegetation Classification Standard in Federal agency activities.

- For the National Vegetation Classification, review the basic statements in the key concepts document.
- Complete the Permanent Housing database, establish a database system between Federal agencies, and improve system communication between Federal and non-Federal vegetation databases.
- Foster international collaboration efforts.

Wetlands Subcommittee

- Continue outreach for the Wetlands Mapping Standard and expand the training course and opportunities.
- Submit the Wetlands Classification Standard proposed maintenance update through the Federal Register for public review and comment.

The Architecture and Technology Working Group

- In the coming year the Working Group will oversee a second year of GeoCloud hosting and facilitation using the two developed reference platform software "stacks" for open-source and commercial geospatial solutions.

Metadata Working Group

- Inform members about ISO metadata and support members in ISO metadata implementation.

FGDC Standards Working Group

- Follow up on findings from the standards use survey. For more information see page 7.
- Identify and recommend ISO, ANSI, and OGC standards for FGDC endorsement in support of the Geospatial Platform.
- Evaluate interconnections among standards bodies and what voting by FGDC member agencies in external standards bodies signifies.
- Review the FGDC standards program of work to identify standards that should be retired or revised and standards projects that should be discontinued.
- Review and recommend the Coastal and Marine Ecological Classification Standard (CMECS) for FGDC endorsement.

User/Historical Data Working Group

- Continue outreach to all interested parties, both inside and outside the Federal Government.
- Continue to provide input on the Geospatial Platform.
- Provide coordinated input to the Library of Congress Geospatial Data Preservation Resource Center.
- Inventory guidance documents on geospatial formats for long-term preservation and access.

APPENDIX A. FGDC LEADERSHIP PROFILES

Anne Castle – *Assistant Secretary for Water and Science, Department of the Interior Chair, FGDC Steering Committee*

Ms. Anne Castle, confirmed as Assistant Secretary for Water and Science in June 2009, is responsible for overseeing water and science policy for the Department of the Interior (DOI), specifically through oversight of the U.S. Bureau of Reclamation and the U.S. Geological Survey. In addition to her leadership on water issues, Ms. Castle has been a champion of the National Land Imaging Program. Prior to joining the DOI, Ms. Castle practiced law for 28 years in Denver, Colorado, focusing on water issues. She received a bachelor of science degree in applied mathematics, with honors, from the University of Colorado and earned a juris doctorate from the University of Colorado.

Dominic Sale – *Acting Deputy Director, Office of E-Government and Information Technology Office of Management and Budget, Vice Chair, FGDC Steering Committee*

Mr. Dominic Sale joined the Office of Management and Budget's Office of E-Government and Information Technology in 2008, working with portfolio management in support of several governmentwide information technology initiatives. Mr. Sale began his Federal career at the Department of Transportation in the Office of the Chief Information Officer. Preceding this, he managed enterprise architecture, capital planning, and security initiatives for various Federal agencies as a contractor for BAE Systems and BearingPoint. Mr. Sale earned a bachelor's degree from the University of Virginia.

Karen Siderelis - *Geographic Information Officer, U.S. Department of the Interior Acting Chair, FGDC Executive Committee*

Ms. Karen Siderelis became the first Geographic Information Officer for the Department of the Interior when she was appointed to the position in September 2008. She previously served as Associate Director for Geospatial Information and Chief Information Officer for the U.S. Geological Survey. Before that, she was Director of the North Carolina Center for Geographic Information and Analysis. Ms. Siderelis earned her master's and bachelor's degrees from the University of Georgia.

Ivan DeLoatch - *Executive Director, Federal Geographic Data Committee*

Mr. Ivan DeLoatch has served as the Executive Director of the FGDC for the past eight years. Previously he served as Chief of the Data Acquisition Branch in the U.S.

Environmental Protection Agency's Office of Environmental Information. He earned a bachelor of science degree in biology from Bowie State University.

Daniel Cotter - ***Chief Technology Officer U.S. Department of Homeland Security***

Mr. Daniel Cotter is responsible for the Department of Homeland Security's enterprise architecture, enterprise data management, and geospatial programs. He holds a master's degree in business administration from Texas A&M University and a master of science degree in geospatial and cartographic sciences from George Mason University. He is a Fellow of the American Association for the Advancement of Science.

Adrian Gardner - ***Chief Information Officer, Goddard Space Flight Center National Aeronautics and Space Administration***

Mr. Adrian Gardner came to the National Aeronautics and Space Administration from the National Weather Service where he had served as Chief Information Officer since January 2007. Mr. Gardner completed a master's degree in public administration at the University of Southern California School of Public Policy and Planning. He also holds a master of science degree in environmental studies from Hood College and a bachelor of science degree in biological science and ecology from the Tuskegee Institute.

Jerry Johnston - ***Geographic Information Officer U.S. Environmental Protection Agency***

Prior to joining the U.S. Environmental Protection Agency in 2005, Dr. Jerry Johnston conducted research on complex environmental systems modeling and served as the project manager for numerous geospatial projects in the private sector. He holds a master's degree and doctorate in environmental science from Indiana University, Bloomington, as well as a bachelor of science degree in environmental science from Michigan State University.

He currently serves as Vice Chair for the National Geospatial Advisory Committee.

Joseph Klimavicz - ***Chief Information Officer and Director, High Performance Computing and Communications, National Oceanic and Atmospheric Administration U.S. Department of Commerce***

Mr. Joseph Klimavicz previously served at the U.S. Department of Defense as the National Geospatial-Intelligence Agency's Deputy Chief Information Officer. He received a bachelor of science degree and a master's degree in engineering from Virginia Polytechnic Institute and State University.

James Kwolek - *Geospatial Information Executive and Director, Office of IT Strategic Planning and Policy Office of the Chief Information Officer, National Geospatial-Intelligence Agency*

Prior to his current assignment, Mr. James Kwolek was the Functional Executive in the Office of Geospatial Intelligence Management at the National GeospatialIntelligence Agency. In 2010 he served as the Associate Deputy Director of Mission Management at the National Counterterrorism Center. He began his career with the Defense Mapping Agency. Mr. Kwolek completed a master of science in technology management from the University of Maryland and received a bachelor of science in geography/cartography from Penn State University.

Stephen Lowe - *Associate Chief Information Officer for Technology Planning, Architecture, and E-Government U.S. Department of Agriculture*

Mr. Stephen Lowe holds graduate degrees in management of information technology from the University of Virginia and in public administration from Virginia Polytechnic Institute and State University. He also holds a bachelor of arts degree in political science from James Madison University.

APPENDIX B. FGDC STRUCTURE AND MEMBERSHIP

The Federal Geographic Data Committee (FGDC) operates under Office of Management and Budget (OMB) Circular A–16 (revised August 2002). The circular incorporates Executive Order 12906 and reaffirms the FGDC's role to provide leadership for the National Spatial Data Infrastructure (NSDI) and to coordinate the development, use, sharing, and dissemination of the Nation's geospatial data. The effective use of geospatial information requires close coordination among the many agencies involved in its development.

The FGDC is an organized structure of Federal geospatial professionals and constituents that provide executive, managerial, and advisory direction and oversight for geospatial decisions and initiatives across the Federal Government. In accordance with OMB Circular A–16, the FGDC is chaired by the Secretary of the Interior or his/ her designee, and the OMB Deputy Director for Management or his/her designee serves as Vice Chair.

FGDC Structure

The FGDC is governed by a Steering Committee that sets the FGDC's high-level strategic direction. The Executive Committee, which is a subset of the Steering Committee, provides advice and guidance to the Chair and the Vice Chair. The National Geospatial Advisory Committee (NGAC) is a Federal advisory committee that provides advice and recommendations on Federal and national geospatial programs. The FGDC Coordination Group conducts the FGDC's day-to-day business. The FGDC Office of the Secretariat, which is located at the U.S. Geological Survey headquarters in Reston, Virginia, provides strategic support and management for FGDC committees, components, and initiatives. The FGDC infrastructure also includes committees, agency-led working groups and subcommittees, and collaborating partners that represent State, Tribal, and local governments, as well as industry and academic and professional organizations. All participants initiate and (or) support the following activities that are crucial to expanding the NSDI and addressing national priorities:

- Providing advice and leadership in applying geospatial capabilities to address national priorities and Presidential initiatives.
- Developing and establishing the National Geospatial Data Clearinghouse on the Internet.
- Developing and implementing standards.
- Creating a national digital geospatial data framework. The framework covers seven fundamental geographic themes (cadastral, elevation, geodetic control, governmental units, hydrography, orthoimagery, and transportation) and is used in concert with many other mission-specific data resources.
- Promoting collaborative relationships for sharing geospatial data among non-Federal partners.
- Developing policies and processes to better harmonize collective action.

More information about the FGDC structure and specific membership can be found at *www.fgdc.gov/participation.*

Steering Committee

The FGDC is governed by the Steering Committee, which is the policy-level interagency group whose central focus is to provide executive leadership for the coordination of Federal geospatial activities between, among, and within agencies. The Committee does this by establishing policy and providing guidance and direction to the member agencies, based on business best practices. The Steering Committee is responsible for overseeing OMB Circular A–16 related activities and for the implementation of the National Spatial Data Infrastructure. The FGDC Chair and Vice Chair lead the Committee, which is made up of senior agency officials for geospatial information (SAOGIs) and includes representatives from Federal organizations, including the Executive Office of the President, the Federal Executive Departments, and independent Federal agencies. A subset of the Steering Committee, the Executive Committee, provides advice and guidance to the FGDC Chair and the Vice Chair on major Federal geospatial priorities and initiatives. The FGDC Chair and Vice Chair lead this committee, which includes representatives from the OMB and the seven Federal agencies that have the largest investments in geospatial technologies. The Executive Committee makes recommendations to the Steering Committee and provides a focal point for coordination with the National Geospatial Advisory Committee.

2011 Steering Committee Agencies

2011 Steering Committee Agencies
Federal Communications Commission (non-voting member)
General Services Administration
Library of Congress
National Aeronautics and Space Administration
National Archives and Records Administration
National Capital Planning Commission (non-voting member)
National Science Foundation
Office of Management and Budget (tie-breaking vote only)
Office of Personnel Management
Small Business Administration
Smithsonian Institution
Social Security Administration
Tennessee Valley Authority
U.S. Agency for International Development
U.S. Army Corps of Engineers (non-voting member)
U.S. Department of Agriculture
U.S. Department of Commerce
U.S. Department of Defense
U.S. Department of Education
U.S. Department of Energy
U.S. Department of Health and Human Services
U.S. Department of Homeland Security
U.S. Department of Housing and Urban Development
U.S. Department of the Interior
U.S. Department of Justice
U.S. Department of Labor
U.S. Department of State
U.S. Department of Transportation
U.S. Department of the Treasury
U.S. Department of Veterans Affairs
U.S. Environmental Protection Agency
U.S. Nuclear Regulatory Commission

APPENDIX C. STATUS OF NSDI DATA THEMES

Office of Management and Budget (OMB) Circular A–16 provides direction for Federal agencies that produce, maintain, or use spatial data in the National Spatial Data Infrastructure (NSDI). The circular identifies 34 data themes of national significance and denotes which are framework data themes. This appendix includes descriptions of the following framework data and data themes: cadastral, elevation, geodetic control, hydrography, orthoimagery, cultural and demographic statistics, soils, watershed boundaries, and wetlands.

Framework Data Theme: Cadastral

Responsible Agency: U.S. Department of the Interior, Bureau of Land Management (BLM)

Description: The BLM is the lead Federal agency for cadastral data and chairs the FGDC Cadastral Subcommittee. Cadastral data describe the geographic extent of rights and interests in real property. Cadastral data include cadastral reference data (see following page) and land parcel data. Under BLM's leadership, the Cadastral Subcommittee develops and implements plans to coordinate cadastral data-related activities among Federal, State, Tribal, and local governments and the private sector and reports on its activities to the FGDC. In the western United States, the FGDC Cadastral Subcommittee continues to work with the wildland fire community to identify sources of land parcel data, standardize available data, and build sustainable operational procedures to provide land parcel data in coordination with States. In other regions of the country, such as the Midwest and the Southeast, the Subcommittee continues to work with State coordinators to identify county cadastral data contacts. The graphic below represents progress as of July 2011. Through September 2011, all of California data and available data in Utah and Wyoming have been standardized.

Parcel Data Standardization Status by County

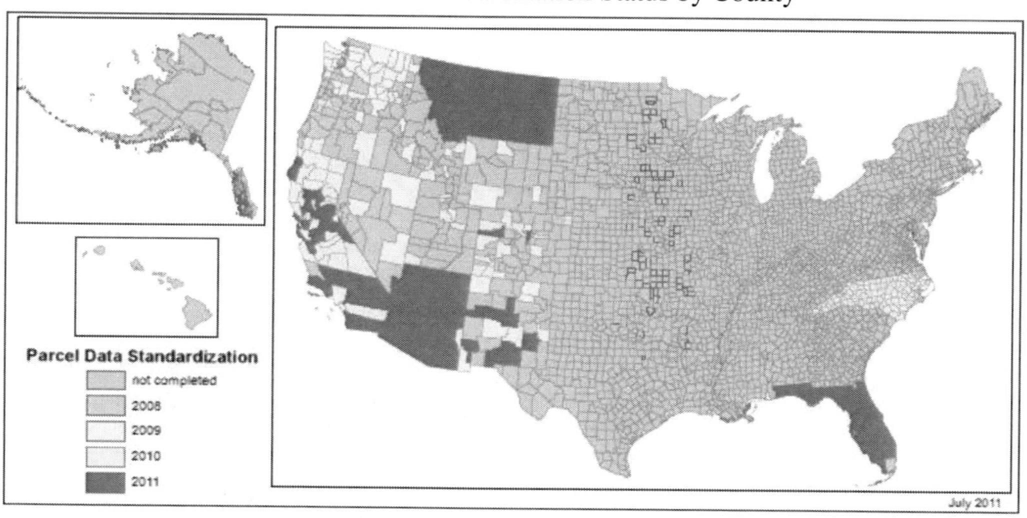

The Public Land Survey System (PLSS) is a cadastral reference system used to divide public domain lands, which are lands managed by the Federal Government.

The original public domain included the land ceded to the Federal Government by the thirteen original States, supplemented with acquisitions from native Indians and foreign powers. It encompasses major portions of the land area of 30 States. Standardized PLSS representation supports geographic information system (GIS) applications that facilitate data integration and enables searches by PLSS location.

The statewide standardization PLSS representation is based on legal cadastral survey records. The graphic below indicates progress through July 2011.

The Cadastral Subcommittee has developed guidance for States to use when developing, publishing, and maintaining all standardized cadastral information (cadastral reference and land parcel data) for government-togovernment data sharing.

State stewardship of cadastral data has seven levels, or phases, that lead to a State aggregating and publishing authoritative local data and serving as a trusted source.

The levels range from level 1, in which a State coordinator identifies and assembles a current inventory of the State's sources of cadastral data, to level 7, in which statewide data are assembled into a single dataset that can be combined seamlessly with other State and Federal cadastral data.

PLSS Standardization Status by State

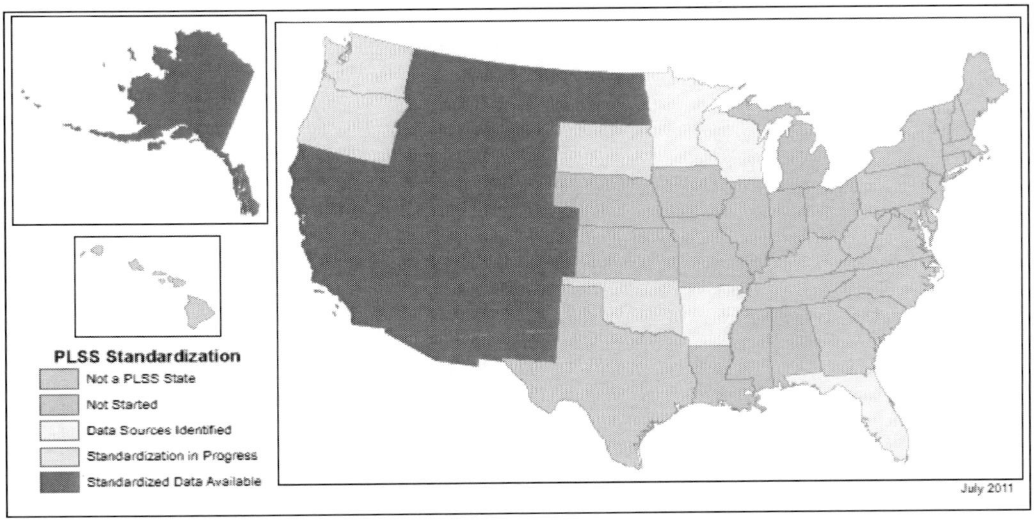

Framework Data Theme: Elevation

Responsible Agency: U.S. Department of the Interior, U.S. Geological Survey (USGS)

Description: The National Elevation Dataset (NED) contains elevation data that provide three-dimensional surface models of the Earth's surface.

The USGS makes elevation data available for land areas and, in cooperation with the National Oceanic and Atmospheric Administration, under coastal waters. The USGS identifies digital elevation data based upon the resolution (spacing between the points) of a grid.

One arc-second resolution (equivalent to 30-meter resolution) elevation data are complete and available for the entire United States, except Alaska. Current USGS efforts are concentrated on providing finer resolution of elevation data at 1/3 and 1/9 arc-second (equivalent to 10- and 3-meter resolutions, respectively) grid spacing. The data are developed from the best available data from Federal, State, and local agencies and private sector partners.

National Elevation Dataset Source Resolution

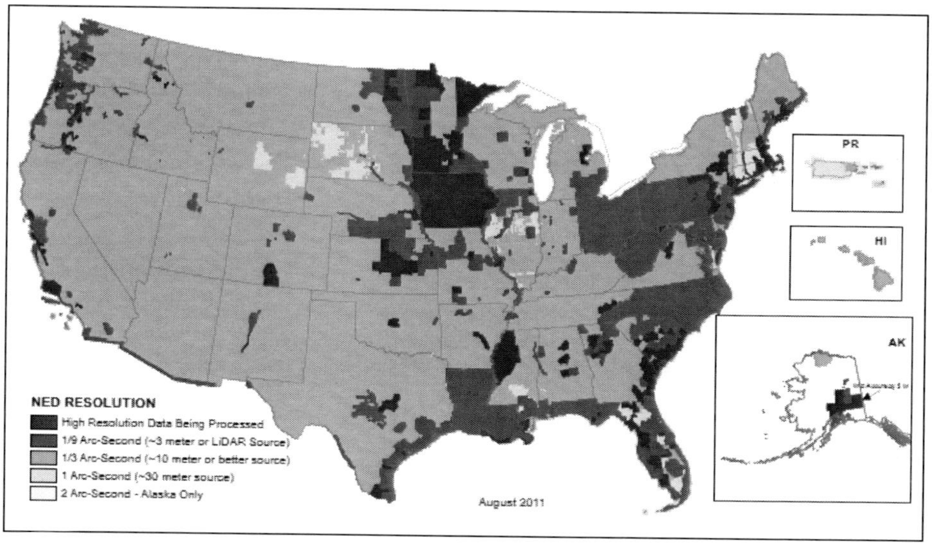

Northeast LiDAR Partners Project

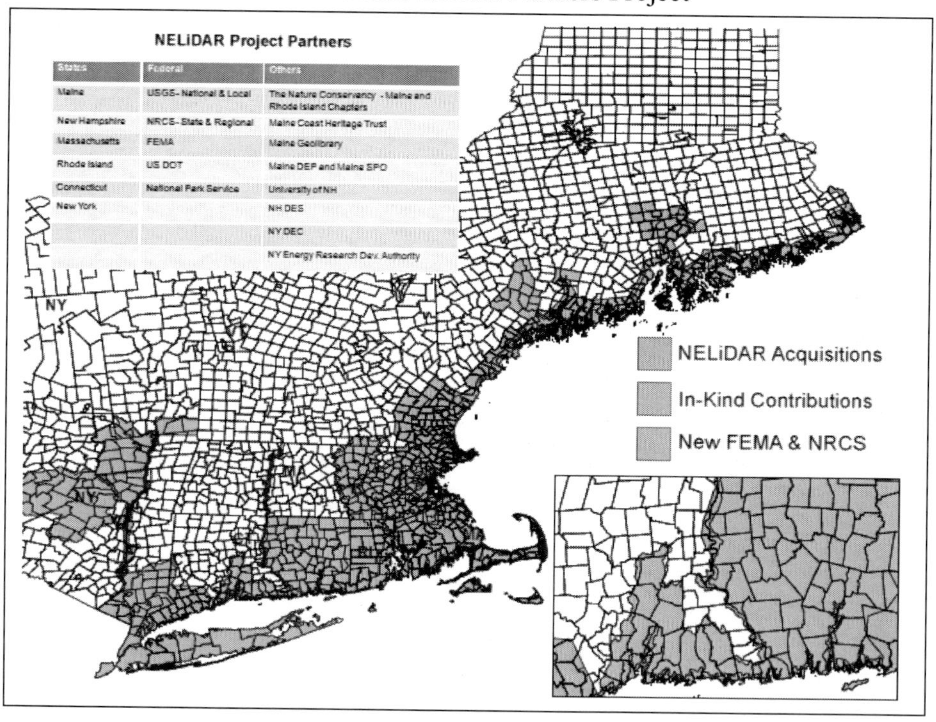

In fiscal year 2011, all high-resolution elevation projects funded through the American Recovery and Reinvestment Act (ARRA) in 2009–10 are to be completed and the data made available for dissemination as Digital Elevation Models from the NED and also, as in the case of LiDAR source data, Log ASCII Standard (LAS) files from the Center for LiDAR Information Coordination and Knowledge (CLICK). Many of these ARRA-funded projects are large, multiple-partner cooperative arrangements requiring precision coordination. One such project is the LiDAR for the Northeast project covering the coastline of New England, New York to the Canada–United States border in Maine.

Six States, five Federal agencies, and eight State agencies and nongovernmental organizations are partners in this project, which will result in approximately 13,500 square miles of new and in-kind LiDAR data in the public domain.

Framework Data Theme: Geodetic Control

Responsible Agency: National Oceanic and Atmospheric Administration (NOAA), National Geodetic Survey

Description: NOAA's National Geodetic Survey manages a network of continuously operating reference stations (CORS) that provide Global Navigation Satellite System data to define and maintain the National Spatial Reference System.

As a multipurpose cooperative endeavor involving more than 200 government, academic, and private organizations, CORS data support three-dimensional positioning, meteorology, space weather, and geophysical applications. Improved coordinates for these geodetic control stations, which are the result of improved analysis of more than 10 years of data, will be made available in fiscal year 2012.

The map shows the distribution of 1,826 stations, including 221 recent additions (green dots).

Continuously Operating Reference Stations (CORS) Network, July 2011

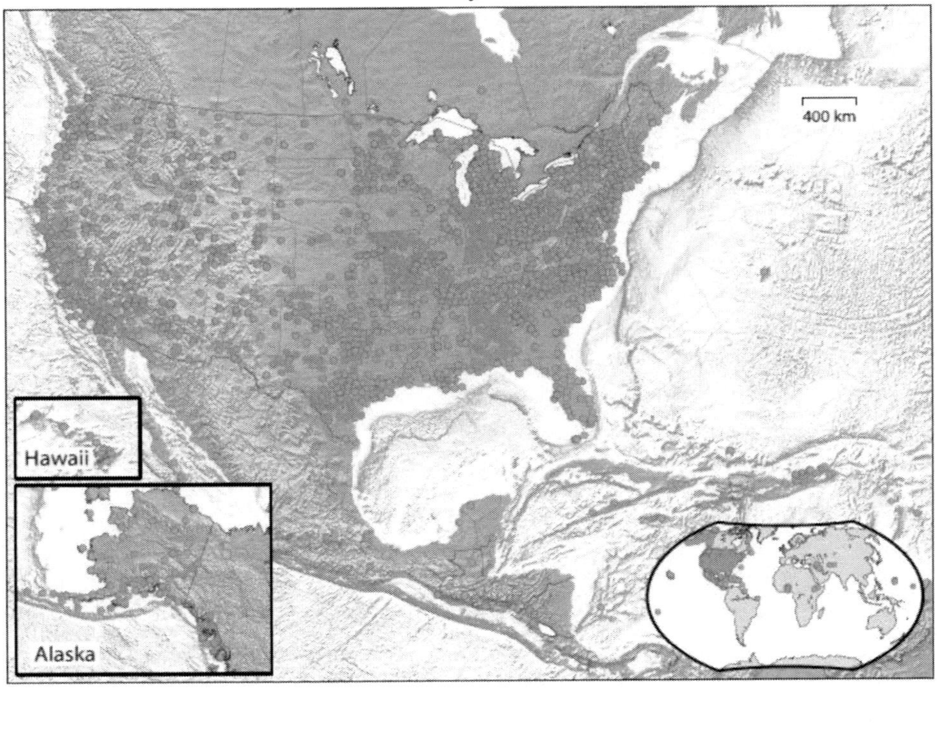

Framework Data Theme: Hydrography

Responsible Agency: U.S. Department of the Interior, U.S. Geological Survey (USGS)

Description: The National Hydrography Dataset (NHD) includes a common data model that contains nationwide coverage of surface water features at 1:100,000 scale and 1:24,000 scale.

These data provide a universal solution for hydrography across the Nation and have widespread application in pollution control, hydrology, resource management, and fisheries research. The USGS provides the central database, technical development, distribution, data integration, leadership, program management, and coordination.

Maintaining these data is largely accomplished through a program of data stewardship using Memorandums of Understanding (MOUs) where the user community is empowered to make necessary updates.

The USGS facilitates the stewardship by providing the methods, tools, training, and guidance to allow users to assume responsibility for maintenance. A complete modernization of the maintenance toolset is underway to expand the user base able to participate in data stewardship.

The USGS provides grant funding to States to help them establish stewardship operations. A number of States are now revising their hydrography data to 1:4,800- and 1:2,400-scale resolutions.

Status of National Hydrography Dataset Stewardship Program

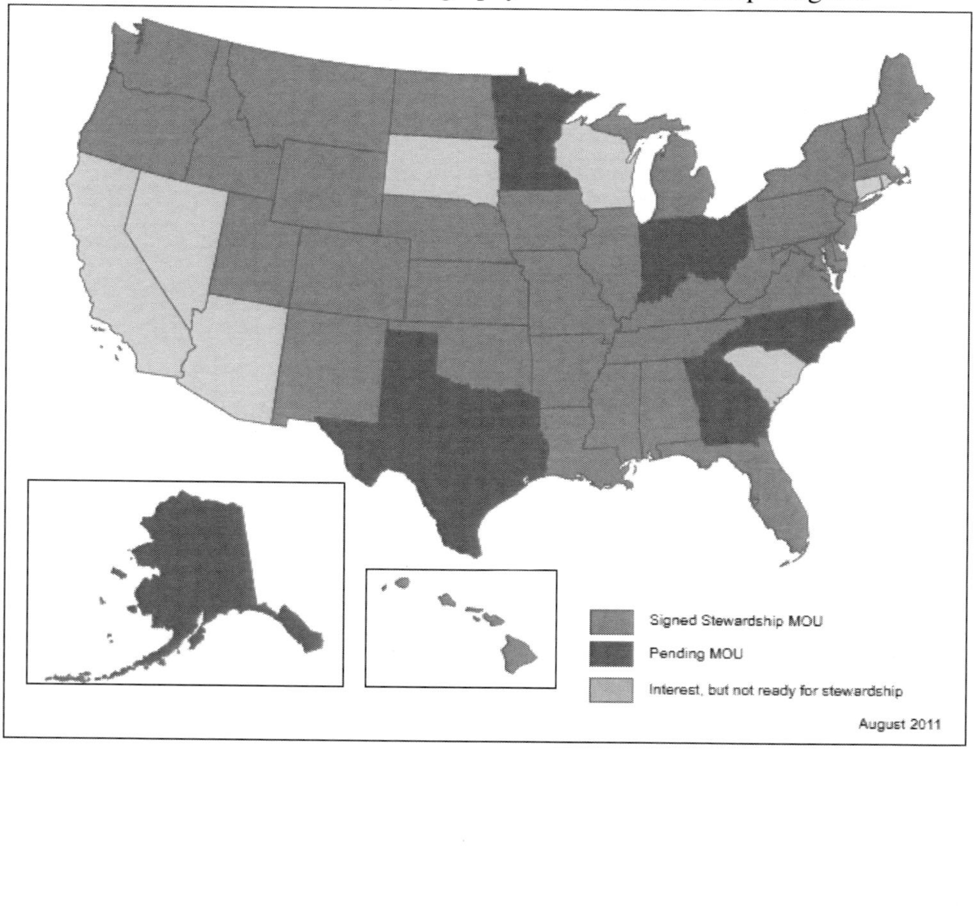

Framework Data Theme: Orthoimagery

Responsible Agency: U.S. Department of the Interior, U.S. Geological Survey (USGS)

Description: The USGS is the designated lead Federal agency for the NSDI orthoimagery data theme and executes its OMB Circular A–16 responsibilities as the theme lead through the National Digital Orthophoto Programs (NDOP).

NDOP was chartered in 1993 as a consortium with the purpose of developing and maintaining national orthoimagery coverage in the public domain by establishing partnerships with Federal, State, Tribal, local, and private organizations.

The consortium members include the Bureau of Land Management, the Farm Service Agency, the Federal Emergency Management Agency, the National GeospatialIntelligence Agency (NGA), the National Oceanic and Atmospheric Administration, the National Resources Conservation Service, the National States Geographic Information Council, the U.S. Census Bureau, the U.S. Forest Service, and the U.S. Geological Survey. The primary Federal programs for NDOP are the Farm Service Agency's National Agriculture Imagery Program (NAIP), the NGA Homeland Security Infrastructure Program (HSIP) 133 Cities Urban Area Initiative, and the USGS National Orthoimagery Program.

USGS National Orthoimagery Program

The USGS National Orthoimagery Program concentrates on acquiring imagery through partnerships with Federal, State, Tribal, regional, and local agencies. The USGS, along with other Federal and State agencies, contributes funds annually to support the USDA-FSA's National Agriculture Imagery Program (NAIP) current maintenance cycle. In addition to direct funding support, the USGS (through the Geospatial Liaisons) coordinates partnerships with State and local agencies to further leverage the NGA's 133 Cities Urban Area Initiative funding pool and expand the project to support local needs. The USGS also provides quality assurance support for the 133 Cities Urban Area Initiative. The USGS makes all public domain digital orthoimagery acquired through its partnership agreements available to the general public through *The National Map* Seamless Server and provides for data archiving. Currently, 1-meter orthoimagery from the NAIP and higher resolution orthoimagery in support of the 133 Cities Urban Area Initiative are available on the Seamless Server. Data availability graphics can be found at *seamless.usgs.gov/data_ availability. php?serviceid= Dataset_5* and *seamless.usgs.gov/data_ availa bility.php? serviceid=Dataset_7*. The USGS National Orthoimagery Program supports the National Geospatial Program by making the imagery available in US Topo maps and through USGS Web services. The USGS National Orthoimagery Program also supports the USGS science programs.

USDA-FSA National Agriculture Imagery Program

The National Agriculture Imagery Program (NAIP) annually acquires imagery during the growing season and is the largest single civilian mapping program in U.S. history. The Farm Service Agency (FSA) and other U.S. Department of Agriculture (USDA) agencies use the imagery to administer farm programs and agriculture-related disaster recovery programs. The imagery is in the public domain and is widely used by Federal, State, and local agencies, as well as by private entities and businesses. A few examples of the business processes that the data have been used to support are economic development, emergency response, environmental management, growth planning, health and human services, homeland security, precision farming and other agribusiness activities, and transportation planning. FSA's priority is to collect data on privately owned farmland as identified by common land unit boundaries, and the agency has established effective cost-share participation with other Federal and State partners for collection outside of those areas. The funding level for NAIP in 2011 supported acquisition of 1-meter imagery for 20 States. Since the program began as a pilot in 2002, it has provided nearly three full cycles of complete 1-meter coverage of the continental United States, giving all States in the lower 48 an imagery currency age of 3 years or less.

2011 NAIP Data Acquisitions

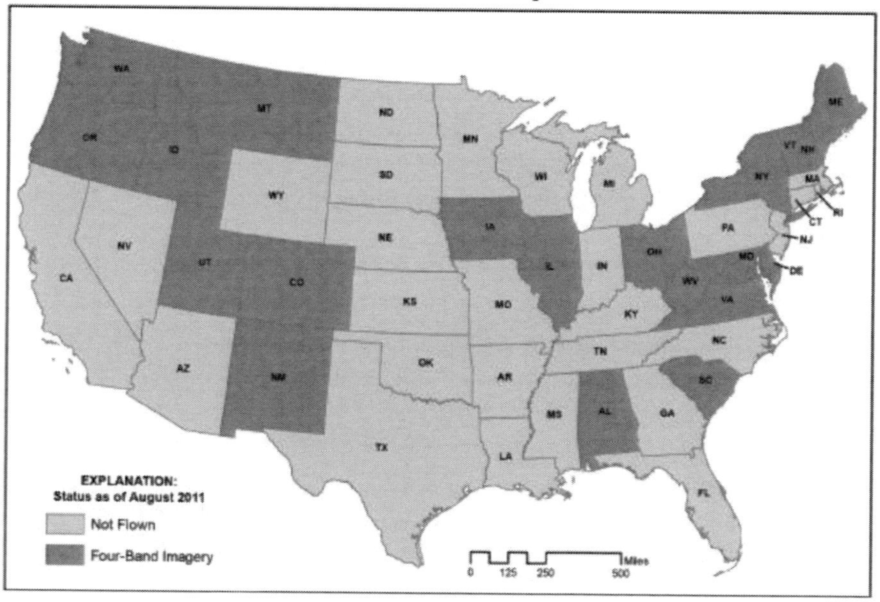

Total Years of NAIP Imagery, 2003–2011 (1- or 2-meter coverage)

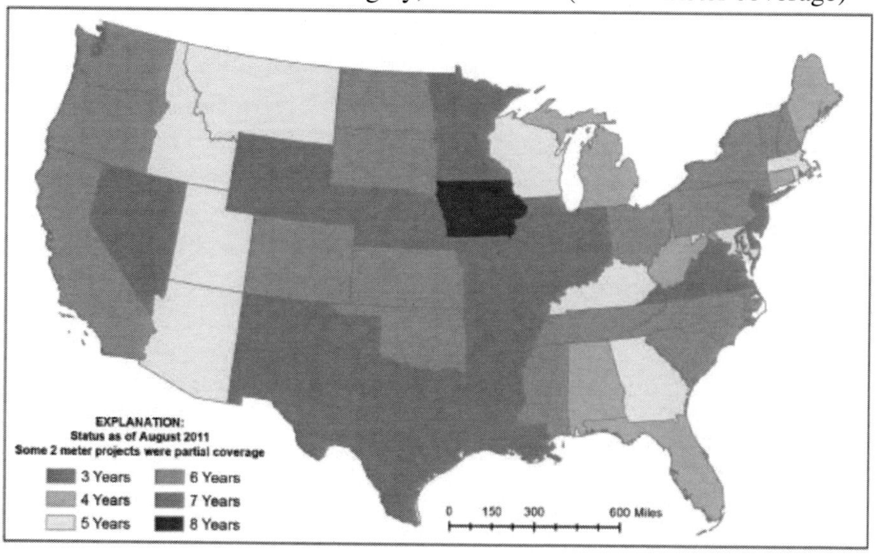

NGA HSIP 133 Cities Urban Area Initiative

The Homeland Security Infrastructure Program (HSIP) 133 Cities Urban Area Initiative annually acquires imagery during the spring and fall leaf-off seasons; the focus of this initiative is the 133 cities that are defined in the 2002 tiger team report that was produced jointly by the National Geospatial-Intelligence Agency (NGA) and the USGS. NGA collects this imagery through partnership with the USGS and uses it to support the Homeland Defense, Homeland Security, and Emergency Preparedness, Response, and Recovery mission to protect the Nation's infrastructure in accordance with Homeland Security Presidential Directive/HSPD–7 and the National Strategy for the Physical Protection of Critical Infrastructure and Key Resources. The collected imagery data also support mutually beneficial partnerships to both enhance and share data among Federal Government agencies and with State and local jurisdictions. This initiative supports partnership and coordination efforts with other Federal agencies such as the Department of Homeland Security, the USGS, and the U.S. Army Corps of Engineers. The collected imagery is largely in the public domain and is widely used by Federal, State, and local agencies, as well as by commercial, private, and nonprofit entities. The USGS provides this imagery to the civilian community through *The National Map* Seamless Server located at the EROS Data Center, and the NGA coordinates imagery distribution to its mission partners. The goal for this initiative is to refresh imagery for the 133 cities on a 2- to 3-year cycle, which has been largely accomplished since the initiative's inception in 2002; imagery for nearly all the cities has been collected twice and collections for many cities are on their fourth or fifth iteration. In fiscal year 2011, the initiative funded 32 projects in 22 States, as well as in the Territory of Guam.

HSIP 133 Cities Urban Area Imagery Initiative:
Fiscal Year 2011 Imagery Acquisitions

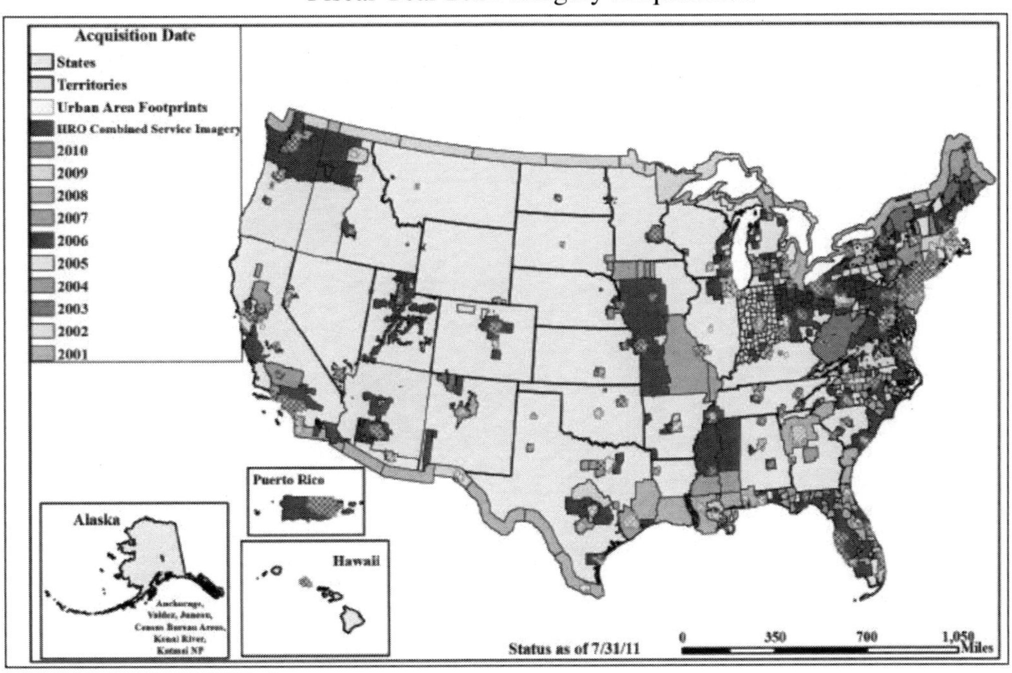

Data Theme: Cultural and Demographic Statistics

Responsible Agency: U.S. Census Bureau

Description: This U.S. Census Bureau-managed theme contains characteristics of people, the nature of the structures in which they live and work, the economic and other activities they pursue, the facilities they use to support their health and recreational and other needs, the environmental consequences of their presence, and the boundaries as well as names and numeric codes of geographic entities used to report the information collected. The U.S. Census Bureau's 2010 Census Profile maps offer a graphic overview of selected demographic information from the 2010 Census of Population and Housing. In addition to a population density map, each page includes a pie chart showing percent of total population by race, a population pyramid, and a bar chart illustrating housing occupancy rates. The map series consists of one page-sized map for each State in the United States, the District of Columbia, and Puerto Rico, as well as a national map. Initial census counts became available in December 2010.

As 2010 census data are tabulated and released, the Census Bureau creates many geospatial products including thematic and reference maps and posts them at: *www.census.gov/ geo/www/.*

As 2010 data continue to be released, the Census Bureau will continue to provide geospatial products reflecting the status of America—changes in population trends and geographic areas—as determined by the 2010 census.

2010 Profile Map United States

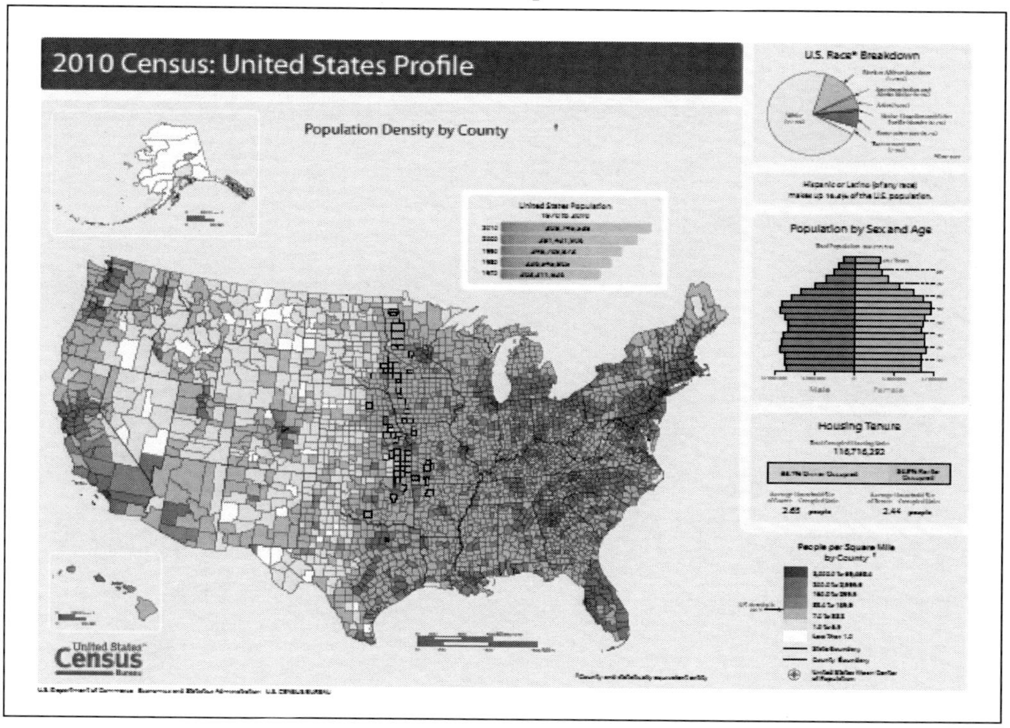

Data Theme: Soils

Responsible Agency: U.S. Department of Agriculture (USDA), Natural Resources Conservation Service Soil Survey Division

Description: The USDA National Resources Conservation Service Soil Survey Division provides leadership and services to produce and deliver science-based soil information (including digital maps, attribute tables, and other types of publications) to help in the understanding and management of global resources.

This service is provided through the National Cooperative Soil Survey (NCSS) program.

The NCSS is a nationwide partnership of Federal, State, and local agencies, and private entities and institutions.

This partnership works collaboratively to investigate, inventory, document, classify, interpret, disseminate, and publish information about soils of the United States and its trust territories and Commonwealths. The activities of the NCSS are carried out on national, regional, and State levels.

The map below illustrates the current status of the Soil Survey Geographic database (SSURGO) for the United States. Dark green color indicates that detailed digital maps and associated attribute tables are available. Light green color indicates that only attribute tables are available. White areas indicate that no soil survey has been completed. During fiscal year 2011, NCSS staff completed 34.8 million acres of initial and update soil survey mapping. In fiscal year 2010, 38.8 million acres of initial and update soil survey mapping were completed.

Available Soil Survey Data, July 21, 2011

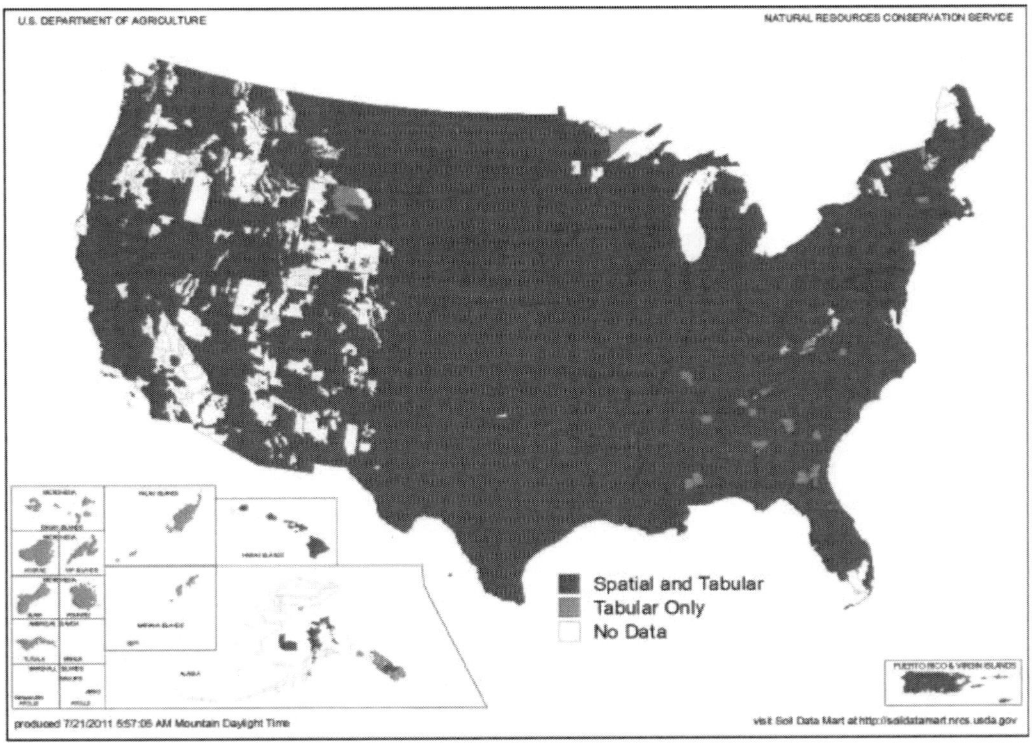

Data Theme: Watershed Boundaries

Responsible Agency: U.S. Department of the Interior, U.S. Geological Survey (USGS), and U.S. Department of Agriculture, Natural Resources Conservation Service (NRCS)

Description: The seamless, national Watershed Boundary Dataset (WBD) is a hydrologic unit framework of uniform size for the United States. It is used for programmatic planning, implementation, and reporting at the national, regional, and local levels. This includes use in multiple analytical and statistical applications. The WBD is certified to national certification standards for the conterminous United States and Hawaii at 1:24,000 scale, the Caribbean at 1:25,000 scale, and Alaska at 1:63,360 scale. Its development and strategic direction is guided under the leadership of the interagency Subcommittee on Spatial Water Data. The USDA-NRCS and USGS have joint management responsibilities under OMB Circular A–16.

Watershed Boundary Dataset (WBD) Status

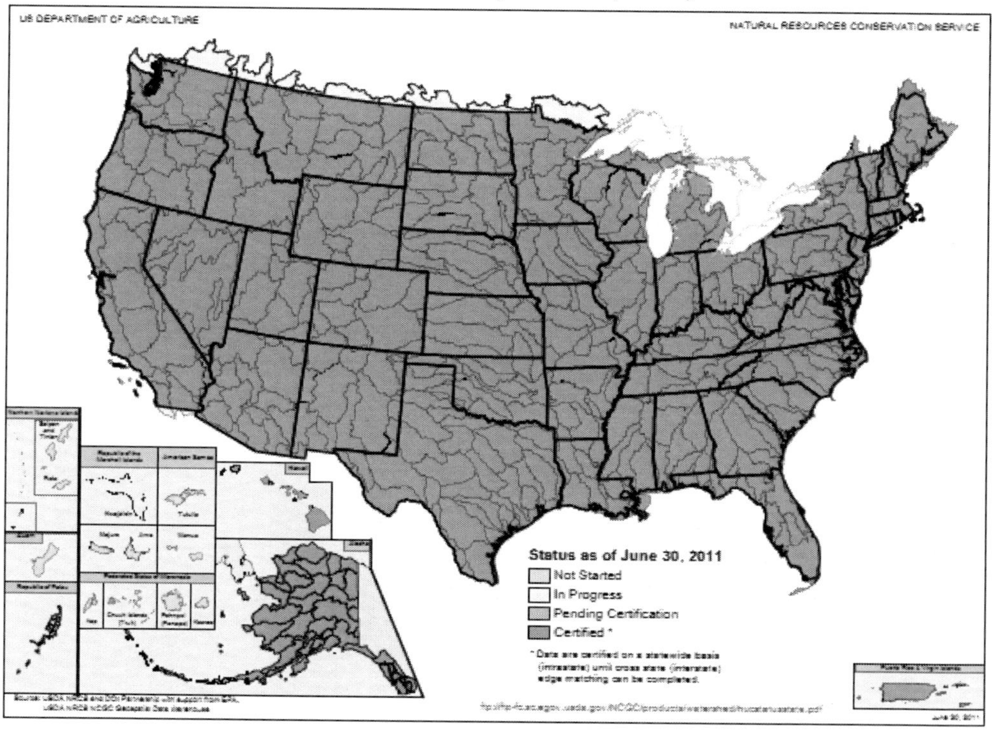

Current efforts are focused on stewardship and integration of the WBD with the National Hydrography Dataset (NHD) and other datasets such as the Geographic Names Information System (GNIS) and harmonization of hydrologic units with Canada and Mexico. Additional data enhancements; for example, improved interpretations along coastal areas, and value-added attribution to facilitate WBD and NHD applications are in progress. WBD planning allows for (but does not require) the incorporation of local partner data at high resolution, plus two additional next down nested tiers at the 14- and 16-digit levels.

This dataset is available through the Geospatial Data Gateway, located at *datagateway.nrcs.usda.gov*

Data Theme: Wetlands

Responsible Agency: U.S. Department of the Interior, Fish and Wildlife Service (FWS)

Description: The National Wetlands Inventory (NWI) provides management, coordination, and delivery of the Wetlands Layer of the NSDI as required under the Emergency Wetlands Resources Act and as part of OMB Circular A–16. The Wetlands Layer consists of wetlands, deepwater, and riparian geospatial data that are served in a seamless geospatial data repository using a state-of-the-art Wetlands Mapper, served through a Web Mapping Service (WMS), downloaded by State, registered through Geospatial One-Stop and Data.gov, and available on the Google Earth™ mapping service. Wetlands data provide the classification, location, and extent of wetlands and deepwater habitats. To facilitate wetlands data development, the FGDC endorsed the Wetlands Mapping Standard in July 2009. The Wetlands Mapping Standard is designed to support current and future digital mapping requirements. It is intended to be extendable, forward-looking, and able to accommodate technology enhancements over time. Adherence to the Wetlands Mapping Standard is required for all wetlands mapping activities funded or conducted by the Federal Government.

The wetlands data layer is 65 percent complete as shown in the status map: 71 percent of the conterminous United States plus another 8 percent in raster scans; 30 percent of Alaska; 100 percent of Hawaii, Guam, and Saipan; and 65 percent of Puerto Rico and the U.S. Virgin Islands. In fiscal year 2011, NWI created a new online Wetland Mapping Training course with training on the Wetlands Mapping Standard and the Wetlands Classification Standard, initiated a Wetlands Subcommittee Working Group maintenance review of the Classification Standard, participated in the Federal GeoCloud effort, and expects to receive over 55 million data requests. The current availability of wetlands data is shown below.

Status of Online Wetlands Data

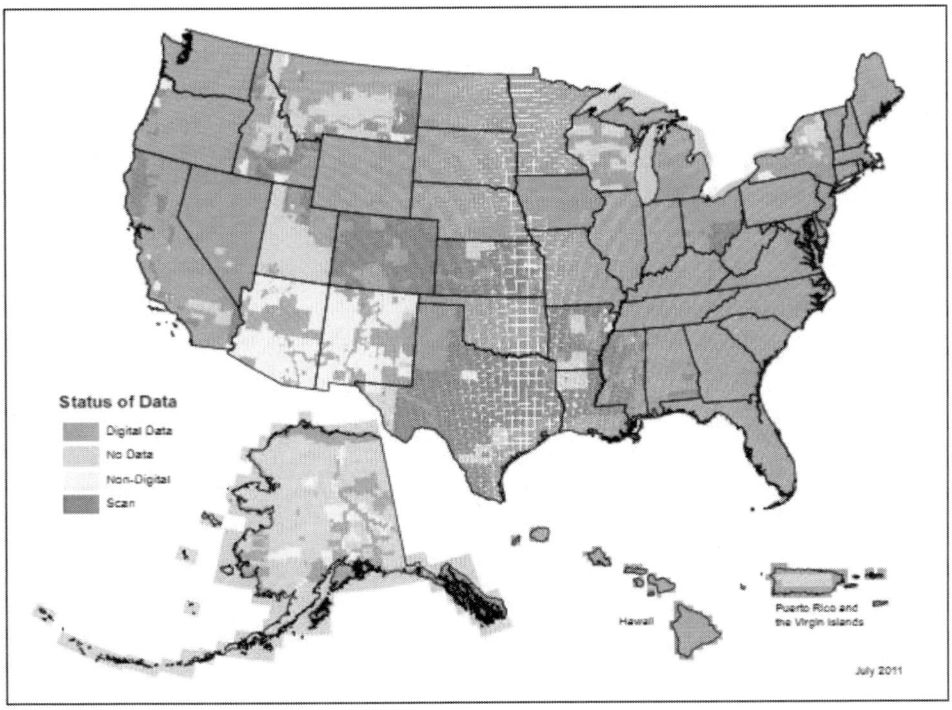

APPENDIX D. GLOSSARY OF ABBREVIATIONS AND TERMS

ANSI	American National Standards Institute
ARRA	American Reinvestment and Recovery Act
ATO	Authority to Operate
BDIA	Broadband Data Improvement Act
CAP	Cooperative Agreements Program
CLDXF	Civic Location Data Exchange Format
CLICK	Center for LiDAR Informaiton Coordination and Knowledge
CMECS	Coastal and Marine Ecological Classification Standard
CORS	Continuously Operating Reference Stations
DMT	Digital Mapping Techniques
DOI	U.S. Department of the Interior
DOT	U.S. Department of Transportation
EPA	U.S. Environmental Protection Agency
FCC	Federal Communications Commission
FEMA	Federal Emergency Management Agency
FGDC	Federal Geographic Data Committee
FSA	Farm Service Agency
FWS	U.S. Fish and Wildlife Service
FY	Fiscal Year
GEO	Group on Earth Observations
Geospatial LoB	Geospatial Line of Business
GEOSS	Global Earth Observation System of Systems
GIS	Geographic Information System
GNIS	Geographic Names Information System
GOS	Geospatial One-Stop
GSA	General Services Administration
GSDI	Global Spatial Data Infrastructure
GWG	Geospatial Intelligence Standards Working Group
HPMS	Highway Performance Monitoring System
HSIP	Homeland Security Infrastructure Program
ISO	International Organization for Standardization
LiDAR	Light Detection and Ranging
MMC	Multipurpose Marine Cadastre
MOU	Memorandum of Understanding
NAIP	National Agriculture Imagery Program
NARA	National Archives and Records Administration
NBII	National Biological Information Infrastructure
NCDDC	National Coastal Data Development Center
NCSS	National Cooperative Soil Survey
NDOP	National Digital Orthophoto Programs (rechartered in 2010 as the National Digital Orthoimagery Program, a subcommittee of the FGDC)
NED	National Elevation Dataset
NENA	National Emergency Number Association
NGA	National Geeospatial-Intelligence Agency

NGAC	National Geospatial Advisory Committee
NGDA	National Geospatial Data Asset
NHD	National Hydrography Dataset
NIMS	National Information Management System
NIST	National Institute of Standards and Technology
NOAA	National Oceanic and Atmospheric Administration
NRCS	Natural Resources Conservation Service
NSF	National Science Foundation
NSDI	National Spatial Data Infrastructure
NSGIC	National States Geographic Information Council
NSRS	National Spatial Reference System
NTIA	National Telecommunications and Information Administration
NWI	National Wetlands Inventory
OGC	Open Geospatial Consortium, Inc
OMB	Office of Management and Budget
PLSS	Public Land Survey System
POC	Point of Contact
RFI	Request for Information
SDI	Spatial Data Infrastructure
SOA	Services-Oriented Architecture
SSURGO	Soil Survey Geographic Database
TDTT	Technical Deployment Task Team
TFTN	Transportation for the Nation
USDA	U.S. Department of Agriculture
USGS	U.S. Geological Survey
USPS	U.S. Postal Service
WBD	Watershed Boundary Dataset
WMS	Web Mapping Service

End Notes

[1] Section 6001(1) of the American Recovery and Reinvestment Act of 2009, Public Law No. 111-5 (2009), codified at 47 U.S.C. 1305(1).

[2] Available at: www.ntia.doc.gov/press-release/2009/ ntia-unveils-program-help-states-map-internetinfrastructure. Accessed: September 5, 2011.

INDEX

C

D

E

F

G

H

I

J

L

M

N

O

P

Q

R

S